Methods in Biotechnology

Edited by
HANS-PETER SCHMAUDER

Editor of the English edition: Michael Schweizer
Translated by: Lilian M. Schweizer

Institute of Food Research,
Norwich, UK

Taylor & Francis
Publishers since 1798

UK Taylor & Francis Ltd, 1 Gunpowder Square, London EC4A 3DE
USA Taylor & Francis Inc., 1900 Frost Road, Suite 101, Bristol, PA 19007

First published 1994 by Gustav Fischer Verlag Jena as *Methoden der Biotechnologie*
hrsg. von Hans-Peter Schmauder

British Library Cataloguing in Publication Data
A catalogue record for this book is available from the British Library.
ISBN 0-7484-0429-5 (cased)
ISBN 0-7484-0430-9 (paperback)

Library of Congress Cataloging in Publication Data are available

Cover design by Youngs Design in Production

Typeset in Times 10/12 pt by Mathematical Composition Setters Ltd, Salisbury,
Wiltshire

Printed in Great Britain by T. J. International Ltd

Contents

[*] Authors listed in Contents have made separate contributions to each chapter

Foreword to English Edition

Biotechnology emerged from developments in the fundamental bio-sciences in the early 1970s to become one of the important new generators of industrial wealth and social benefit. The products of this technology have enabled the early detection of diseases, and provided cures for previously incurable critical illnesses. They have given industry new ways to make valuable chemicals and to treat its waste. The technology has influenced what we eat, how we wash our clothes and even how we solve crimes.

Looking to the future, current discoveries in the bio-sciences will lead to new commercial opportunities. The improved understanding of genome structure–function relationships and developmental genetics which will emerge from research programmes in integrative biology will impact upon healthcare and industry. In medicine, this will enable critical genetic diseases to be treated by therapies which rectify gene abnormalities in diseased tissues. The development of such *in vivo* gene expression vectors poses new problems for biotechnologists and bioengineers, since their production and purification characteristics are very different from conventional protein therapeutics. New ways must be found to produce supra-molecular gene constructs and viral expression systems. The targeting of such therapeutics to specific organs in the body is also problematic yet amenable to scientific and engineering analysis. For industry, developments in bioinformatics and the ability artificially to accelerate genome evolution will lead to commercially valuable, *de novo* designed enzymes and proteins with novel functions.

This growth of product opportunity has demanded the development of new experimental methods, and the multidisciplinary effort of fundamental scientists and engineers for its commercial exploitation. Critical to the future success of biotechnology is the development of teaching materials that will enable undergraduates to acquire the new skills and experimental techniques on which it is based. *Methods in Biotechnology* promotes this by providing a strong foundation for the development of effective university practical courses in the discipline. I hope that it contributes to the successful training of a new generation of biotechnologists who are well equipped to exploit the possible potentials.

Professor Nigel K. H. Slater
Aston University

Foreword to German Edition

The field of biotechnology has been the scene of many stormy developments in the past decade. In some universities biotechnology is a degree course in its own right, and in others it has a place alongside microbiology/biology, biochemistry/chemistry and engineering sciences/process technology, the basic disciplines of biotechnology. A good methods book is essential for these courses. Although there have been many excellent books written about biotechnology, a text describing the methods used and which is suitable for university and college students or for scientists taking up the subject has been lacking. Such a book has now been edited by Professor Schmauder.

This book demonstrates the necessity of acquiring a thorough grounding in basic techniques before going on to carry out experiments in biotechnology, and it emphasises the importance of the interdisciplinary nature of biotechnology.

There can be no doubt that the book *Methods in Biotechnology* will encourage teachers in universities and other institutes of higher education to take up this subject. It is to be hoped that modern biotechnology, in Germany often irrationally neglected, will receive more support, given that we may assume that in the coming decade over ninety per cent of conventional processes will be carried out with genetically manipulated organisms. This would create new jobs and considerably improve the employment opportunities for students from many disciplines. I would like to wish this book every success.

Hans-Jürgen Rehm
Münster

Preface

This book is unusual in that it combines suggestions for experiments in classical microbiology, fermentation and tissue culture with those for biochemical and molecular biological investigations. Its specific aim is to emphasise how much basic science is required for biotechnology. With the help of this book undergraduates can experience biotechnology at the lab bench. Special attention has been paid to correct the text, but in no way has the original format been altered. The English translation of H.-P. Schmauder's book should make this very informative text available to a wider audience.

Lilian M. Schweizer
Michael Schweizer

Introduction

H.-P. SCHMAUDER

Biotechnology is an interdisciplinary subject; its practice requires cooperation between the scientist, the engineer and the economist, combined with an understanding of problems which can arise in the different fields. The approaches taken by each of the specialists in dealing with the problems obviously vary, and this entails that an appreciation of, and a willingness to learn, techniques both simple and complicated in an unfamiliar field are essential.

The aim of this book is to bridge the gaps and to improve the cooperation between the above-mentioned disciplines. For this reason the descriptions of the techniques and methods all follow the same format:

1 Title
2 Basic theory and objective
3 Description of the experiment and procedure subdivided as follows:
   (i) Principle of the experiment
   (ii) Strains, media and equipment required
   (iii) Experimental procedure

> (a) Interpretation, presentation of results, conclusions (e.g. with suggestions for further experiments)
> (b) Literature references (as a starting point for more detailed study of the topic).

Information is provided on safety, protective measures, taxonomy of individual microorganisms and techniques, as well as economic questions in the relevant chapters whenever necessary. Although these issues should always be a main consideration for all scientists and others interested in biotechnology, a detailed discussion of these topics is beyond the scope of this book. The importance of patents and strain collection is, however, worthy of mention.

Patents of biotechnological relevance are differentiated according to whether they are object or product patents, method patents or patents for the innovative use of novel or known compounds or pathways. Product patents include those for

Table 1.1 Major culture collections.

1 General information is available from:

- WFCC World Data Center, Riken, 2-1 Hirosawa, Wako, Saitama 351-01, Japan
- Dr R. Kokke, WFCC Treasurer, CBS, P.O. Box 273, Oosterstraat 1, NL-3740 AG Baarn, The Netherlands
- European Culture Collections Organization (ECCO), Sekretariat des Informations-zentrums, Mascheroder Weg 1b, D-38124 Braunschweig, Germany

2 CCM, Czechoslovak Collection of Microorganisms, Masaryk University, Tvrdeho 14, CZ-602 00 Brno, Czech Republic

3 DSM, Deutsche Sammlung von Mikroorganismen und Zellkulturen GmbH, Mascher-oder Weg 1b, D-38124 Braunschweig, Germany (Plant cells are also available)

4 CBS, Centraalbureau voor Schimmelcultures, P.O. Box 273, Oosterstraat 1, NL-3740 AG Baarn, The Netherlands

5 CCAP, Culture collection of algae and protozoa, Institute of Freshwater Ecology, The Windermere Laboratory, Far Sawrey, Ambleside, Cumbria, LA22 0LP, UK

6 ECACC, European Collection of Animal Cell Cultures, PHLS Centre for Applied Microbiology and Research, Porton Down, Salisbury, Wiltshire, SP4 0JG, UK

7 IMI, International Mycological Institute, Genetic Resources and External Services, Bakeham Lane, Engelfield Green, Egham, Surrey, TW20 9TY, UK

8 NCFB (NCDO), National Collection of Food Bacteria, Institute of Food Research, Earley Gate, Whiteknights Road, Reading, RG6 2EF, UK

9 NCYC, National Collection of Yeast Cultures, Institute of Food Research, Norwich Research Park, Colney, Norwich, NR4 7UA, UK

10 ATCC, American Type Culture Collection, 12301, Parklawn Drive, Rockville, Maryland 20852, USA

11 NRRL, Agricultural Research Service Culture Collection, Northern Regional Research Center ARS/USDA, 1851 North University Street, Peoria, Illinois 61604, USA

12 IfO, Institute for Fermentation, Osaka, 17-85 Juso-Honmachi 2-chome, Yodogawa-ku, Osaka 532, Japan

13 UQM, Department of Microbiology, University of Queensland, St Lucia, Queensland 4067, Australia

14 Pilzkulturensammlung, Friedrich-Schiller University Jena, Freiherr von Stein Allee, D-99425 Weimar, Germany

15 Hans-Knöll-Institut für Naturstoff-Forschung e. V., Beutenbergstr. 11, D-07745 Jena, Germany

16 Seeds, but not plant tissue cultures can be obtained from national and international gene banks. Information on the international gene banks (CIP, IRRI, etc.) is available from:

- IBPGR, International Board for Plant Genetic Resources, c/o FAO of the United Nations, Via delle Sette Chiese 142, I-00145 Roma, Italy
- Institut für Pflanzengenetik und Kulturpflanzenforschung - Genbank, Corrensstr. 3, D-06466 Gatersleben, Germany
- Institut für Pflanzenbau und Pflanzenzüchtung, Bundesanstalt für Landwirtschaft, Bundesallee 50, D-38116 Braunschweig, Germany

chemically defined substances, apparatus and recommendations. This type of patent affords protection of production, use, advertising and selling. Method patents protect production, methods and techniques. The latter category covers the production of biomass, interesting compounds, etc. by microorganisms. In most countries it is not possible to patent an organism (Crespi, 1988). Major culture collections are listed in Table 1.1.

A prerequisite for patent protection of procedures using microorganisms is that the strain used is deposited in an internationally recognised culture collection. Depositing important strains in the strain collections is also to be recommended as a safety measure. The culture collections offer assistance when a certain strain is required for a special method or a specific experiment. Any non-patented strain listed in the catalogue can be purchased.

Reference

CRESPI, R.S. (1988) *Patents: a basic guide to patenting in biotechnology* (Cambridge: Cambridge University Press).

Basic Scientific Techniques for Biotechnology

Biotechnology is an interdisciplinary topic and, as such, requires that practitioners have knowledge of basic techniques from the constituent disciplines. There are several reference works dealing with the different techniques; however, it has been decided to mention some of the routine methods here so that the experiments described in the book can be carried out without reference to numerous other works.

2.1 ANALYTICAL METHODS

A.W. Alfermann, K. Dombrowski, M. Petersen, H.-P. Schmauder and M. Schweizer

All measurements and controls should be carried out in duplicate and, if practicable, in triplicate.

2.1.1 *Determination of Inorganic Compounds*

Ammonium

This method was originally described by Fawcett and Scott (1960) and is based on the complex reaction between ammonium ions and hypochlorite, phenol and nitroprusside to form a blue indophenol anion.

REAGENTS

Ammonium standard: 44.4 mg NH_4NO_3 in 1000 ml dH_2O (10 $\mu g/ml$; OD 0.95). *Phenol reagent*: 25 g phenol; 312 ml 1 N NaOH, dH_2O ad 1000 ml. *Sodium nitroprusside reagent*: 1 g nitroprusside sodium, 100 ml dH_2O (stable for one month at 4 °C). *Sodium hypochlorite solution*: 22 ml NaOCl solution (approx. 15% active chlorine), 478 ml dH_2O; if the NaOCl solution is not fresh, the volume added should be increased.

Dilute the sodium nitroprusside solution $1:100$ in dH_2O. Prepare ammonium standards by setting up a series of dilutions of the ammonium nitrate solution in the range $1-5$ ml made up to 10 ml in dH_2O.

- sample (H_2O for the reference)   1 ml
- dH_2O 1 ml
- phenol reagent   2 ml
- sodium nitroprusside solution   3 ml
- sodium hypochlorite solution   3 ml

Shake well and incubate for 30 min in the dark at room temperature. Determine OD_{630} (1 ml cuvette).

Nitrate

Nitrate is a nitrogen source which often regulates growth rate, and its concentration is most conveniently measured by an optical test (*Method A*). This procedure is based on observations made by Hoather and Rackham (1959) and involves absorption measurements at 210 nm and 275 nm. An alternative method for determining the concentration of nitrate is that ascribed to Sawicki and Scaringelli (1971) (*Method B*) in which nitrate is reduced to nitrite; nitrite reacts with a diazo reagent to give a coloured product.

Method A

ddH_2O. *Nitrate stock solution*: 0.7221 g KNO_3, ddH_2O ad 1000 ml (1 ml = 0.100 mg N = 0.443 mg NO_3). *Nitrate standard*: 100 ml nitrate stock solution made up to 1 l with ddH_2O (1 ml = 10 μg N = 44.3 μg NO_3).

Calibration measurements are made with the standard solution in the range 0 to approx. 75 mg NO_3/l (0–20 mg N/l). The absorption (OD) at 210 nm and 275 nm of the standard and the unknown samples is measured using either 1, 2 or 4 cm cuvettes. All readings are measured against ddH_2O.

All OD measurements are converted using the following formula:

$$OD = OD_{210} - (4 \times OD_{275})$$

The nitrate content of the unknown samples is determined from the standard curve obtained with the values of the standard samples. A possible source of error is high concentrations of organic material. All samples should be filtered (membrane or paper) since turbidity can also affect the results.

Method B

REAGENTS

Copper sulphate solution: 3.94 mM $CuSO_4 \times 5 H_2O$. *Sodium hydroxide*: 0.1 N NaOH. *Hydrazine sulphate solution*: 0.21% (w/v) in ddH_2O. *Acetone*: 10% (v/v) in ddH_2O. *Diazo reagent*: 10% (v/v) ortho-phosphoric acid (85%), 4% (w/v) sulphanilamide, 0.2% (w/v) N-naphthyl-(1)-ethylene-diammonium dichloride in ddH_2O. *Nitrate standards*: 5, 10, 20, 30 mM KNO_3 (0.505, 1.01, 2.02, 3.03 g KNO_3/l).

PROCEDURE

The samples must be diluted with ddH_2O.

- probe (diluted)   4.0 ml
- copper sulphate solution 0.2 ml
- 0.1 N NaOH   1.0 ml
- hydrazine sulphate   0.2 ml

The samples are mixed and incubated in a water bath at 65 °C for 30 min. To prevent evaporation the test tubes are 'stoppered' with glass marbles. At the end of the incubation the tubes are placed on ice for 5 min before the addition of:

- 10% acetone   0.4 ml
- diazo reagent 1.2 ml
- ddH_2O 3.0 ml

Mix thoroughly and allow to stand at room temperature for 15 min. Measure OD_{540} of the sample against a blank containing nitrate-free medium or water. A standard curve is prepared using the standard solutions of 5, 10, 20 and 30 mM KNO_3 treated in the same way as the samples.

Phosphate

Inorganic phosphate or organically bound phosphate can be determined either colorimetrically or enzymatically. In biotechnological processes the determination of inorganic phosphate as a substrate is important because phosphate can have a regulatory function (enhancement of growth, in the biosynthesis of secondary metabolites by inhibiting product formation). The routine methods for determining the amount of inorganic phosphate are based on the formation of molybdane blue. The technique developed by Gomori (1942) (*Method A*) and the older method of Martland and Robinson (1926, cited in Kleber *et al.*, 1990) (*Method B*) are those most commonly used. The determination of organic bound phosphate following enzymatic removal of the phosphate group by acid or alkaline phosphatase or in an enzyme-coupled test is also based on these methods.

Method A

REAGENTS

Sulphuric acid ammonium molybdate solution: 25 ml 10 N H_2SO_4, 50 ml 2.5% $(NH_4)_6Mo_7O_{24}$, 325 ml ddH_2O. *Reducing solution*: 40 g sodium disulphate, 1 g

Photorex (Merck), ddH$_2$O ad 100 ml. *Phosphate standard*: 1.5 mM NaH$_2$PO$_4$ (17.99 mg/100 ml).

Warning! for both methods it is essential that glassware and plastic equipment are rinsed several times with distilled water because all detergents contain phosphates.

PROCEDURE

- medium (or phosphate-free blank or standard)   1.0 ml
- ammonium molybdate solution   4.0 ml
- reducing solution   0.5 ml
- ddH$_2$O   4.5 ml

Mix thoroughly and allow to stand at room temperature for 30 min before determining OD$_{660}$ using a cuvette with a 1 cm light path.

Method B

REAGENTS

Standard phosphate solution: 0.2 g KH$_2$PO$_4$ in 1000 ml dH$_2$O. *Ammonium molybdate*: (a) 5 g ammonium molybdate in 40 ml dH$_2$O; (b) 17 ml concentrated sulphuric acid diluted with 40 ml dH$_2$O. Immediately prior to use the solutions (a) and (b) are mixed and brought up to 100 ml with dH$_2$O. *Hydroquinone/sodium bisulphite*: (a) 0.5 g hydroquinone in 40 ml dH$_2$O; (b) 20 g sodium bisulphite in 40 ml dH$_2$O. Immediately prior to use the solutions (a) and (b) are mixed and made up to 100 ml with dH$_2$O.

PROCEDURE

- sample (or water)   0.1 ml
- ammonium molybdate solution   2.0 ml
- addition of hydroquinone / Na bisulphite solution 1 min later   1.0 ml
- dH$_2$0   2.5 ml

Leave to stand at room temperature for 30 min before determining OD$_{720}$ (1 cm cuvette). For the calibration curve the phosphate standard solution should be diluted 1:1 to 1:8.

2.1.2 *Determination of Organic Compounds*

Glucose/Sucrose

Several methods have been described for the determination of glucose or sucrose: e.g. refractometry, chemical approaches using Anthron or other reagents, as well as HPLC. The most useful method is dependent on glucose oxidase, because it allows a very accurate and specific determination of the substrate, glucose. It is not possible to give an exact description of the method which is valid for all purposes, because each test system is dependent on the source of the glucose oxidase and on the enzyme to which it is coupled to give the colour reaction. (Glucose detection kits are available from Boehringer Mannheim.) The amount of sucrose can be determined by first hydrolysing the disaccharide into glucose and fructose with 1 N hydrochloric acid (or sulphuric acid) at 60 °C, neutralising the sample and then determining the amount of glucose. (Do not forget to take the dilutions caused by the hydrolysis and the neutralisation into account when calculating the final result!) A calibration curve should be drawn up for every experiment.

HPLC measurements with commercially available enzyme-based glucose sensors are used routinely to determine glucose concentrations.

REFRACTOMETRY

An Abbé refractometer is required for this method. The decrease in the amount of sugar in the culture medium can be followed using a refractometer. An aliquot of the medium is pipetted onto the refractometer, and the refractive index is determined. The disadvantage of this method is that other refractile substances in the medium can influence the refractive index, and therefore the value obtained should only be regarded as an approximate one. For many microorganisms it is known that the sucrose added to the media is split into glucose and fructose and that the glucose is used up more quickly than the fructose. Obviously this differential use of the two sugars cannot be measured using a refractometer. For such kinetic measurements either a combination test system, from Boehringer Mannheim, or HPLC measurement should be adopted.

Amino Acids

The amount of free amino acids in the medium and in the cell is an indication of the physiological state of the cell. Free amino acids can be determined with the help of auxotrophic bacteria. The bacteria-based method also allows a determination of specific amino acids. The strains are available from the various culture collections, and in general they have a very sensitive response to the specific amino acids. If it is necessary to know only whether amino acids are present or not then the ninhydrine method using one amino acid (e.g. L-leucine) as the standard can be used.

If the intracellular concentration of amino acids is to be determined, the method described by Roos (1989) for the filamentous fungus *Penicillium cyclopium* and the following method for plant tissue can be used. Mycelia or plant cells are incubated under shaking with 5 ml 90% EtOH, for 3–4 h, centrifuged and washed with 80% EtOH. The ethanol phases are combined and used for amino acid determination. With this method the lower molecular weight and more soluble amino acids are separated from proteins and peptides.

REAGENTS FOR AMINO ACID DETERMINATION

L-leucine: 250 μg/ml. *Buffer*: 272 g sodium acetate × 3 H$_2$O, 50 ml glacial acetic acid made up to 500 ml with dH$_2$O, pH 5.5. *Ninhydrine solution*: dissolve 2 g in 50 ml purified methyl glycol and make up to 100 ml with buffer. *Ascorbic acid solution*: 0.1% in dH$_2$O.

PROCEDURE

- sample in ethanol (or ethanol for the control)   0.2 ml
- ninhydrine solution   0.5 ml
- ascorbic acid solution   0.05 ml

The samples are placed in a water bath, kept at boiling point for 15 min, then cooled down quickly by holding under running water before adding 5 ml 50% EtOH. Measure OD_{575} of the samples using a 1 cm cuvette. The standard values are obtained using $15-250$ $\mu g/ml$ L-leucine.

Proteins

There are numerous descriptions of methods for determining protein concentration. All methods depend on specific reactions of individual linkages or specific constituents of proteins: e.g. peptide bonds, the aromatic hydroxyl group of tyrosine, etc. Depending on the method and the protein used as the standard, varying degrees of sensitivity are obtained; therefore the method of protein determination should not be changed within a series of experiments. In particular, when measuring proteins extracted from cells and when using certain buffer systems, e.g. TRIS, or enzyme stabilisers, e.g. glycerol, the effects of these substances on the protein measurement method have to be checked. This is generally done by performing the test using the buffer with and without the addition of the stabilizing substances. Generally it is sufficient to include the buffer and additives when making the calibration curve. The most widely used protein determination methods are based on those originally developed by Lowry *et al.* (1951) and Bradford (1976). Both methods will be described here. The standard used in each system is either human or bovine serum albumin. Two to four control tubes should be included in every series of measurements. If the protein content of cells is to be determined, the cells can be extracted with sodium dodecyl sulphate (SDS).

Extraction of Protein from Cells

The precipitate obtained in the 80% EtOH extraction used for the amino acid determination can be used to determine protein content, or a fresh aliquot of cells can be extracted.

The residue from the ethanol extraction is taken up in 2% SDS and the proteins are extracted by shaking O/N (foaming occurs!). Cell debris is removed by centrifugation and the supernatant can be treated with trichloroacetic acid (TCA) (see below).

Fresh cells are ground with sand and taken up in buffer to give a suspension from which the proteins can be precipitated by incubation in an ice bath with 10% TCA. The samples are left to stand on ice for a further 15 min before transferring to a 90 °C water bath for 15 min. The samples are centrifuged and the precipitate is washed with 0.5% TCA. The precipitate which contains the proteins is taken up in 10 ml 1 N NaOH and left to stand O/N before the protein content is determined, e.g. according to Lowry *et al.*

Method According to Lowry et al.

This method is based on the molybdane blue reaction. The test is very sensitive, and protein concentrations between 5 and 80 μg can be measured.

Sodium carbonate: 2% in ddH$_2$O (must be made up freshly each time). *Copper sulphate*: 1% in ddH$_2$O. *Sodium potassium tartrate*: 2% in ddH$_2$O. The copper sulphate and sodium potassium tartrate solutions are mixed in equal proportions before use. *Sodium carbonate / copper sulphate / tartrate solution*: 49 parts sodium carbonate solution plus 1 part of the copper sulphate / sodium potassium tartrate solution. *Folin phenol reagent* is commercially available.

PROCEDURE

- alkaline protein extract (or 1 N NaOH)   0.3 ml
- H$_2$O (or aqueous protein solution)   0.5 ml
- sodium carbonate / copper sulphate / tartrate solution   3.0 ml
- Allow to stand for 10 min
- Folin phenol reagent   0.3 ml

Mix thoroughly and leave to stand for 45 min before measuring OD$_{578}$ using a 1 cm cuvette.

Method According to Bradford

This method relies on the binding of the dye Coomassie-Brilliant blue G250 to proteins and has been developed from the staining of proteins in gels. The method is both quick and sensitive; the linear range is between 1 and 100 μg.

REAGENTS

Dissolve 100 mg Coomassie-Brilliant blue G250 in 50 ml 95% EtOH, add 100 ml 85% phosphoric acid and make up to 600 ml with dH$_2$O. The solution is filtered and 100 ml glycerol is added, then the solution is made up to 1000 ml. The solution can be used 24 h later. The solution, which is stable for at least 4 weeks, must be stored in a brown bottle.

PROCEDURE

- protein solution (standard or extract)   0.1–1.0 ml
- Bradford's reagent   5.0 ml

(If the protein concentration is low, it is recommended that only 1 ml of Bradford's reagent be used.) Allow the samples to stand at room temperature for 10–30 min before reading OD$_{595}$ against a blank consisting of buffer plus Bradford's reagent.

Fats and Lipids

Highly specialised analytical methods are necessary for the determination of fats and lipids, and they have to be adapted for the organism or process involved because interference is possible from any number of sources. Gravimetric methods and HPLC are the techniques generally favoured.

DNA, RNA

The best method for determining DNA and RNA is spectrophotometry of the dissolved nucleic acids according to the following rules.

Concentration of plasmid or genomic DNA:

$$OD_{260} \times 50 = \mu g/ml$$

Concentration of RNA or single-stranded DNA:

$$OD_{260} \times 36 = \mu g/ml$$

Concentration of oligonucleotides:

$$OD_{260} \times 25 = \mu g/ml$$

To obtain a rough estimate of the amount of protein in nucleic acid preparations the extinction of the solution at 280 nm is determined. If the quotient of the extinctions at 260 nm and 280 nm lies between 1.8 and 2.0 the solution is regarded as being free from protein.

Detailed descriptions of the methods for determining these important macro-molecules can be found in Section 2.3 of this book.

2.1.3   *Growth and Cell Viability*

The amount of biomass produced as well as the production of intermediates and their conversion is, to a large extent, determined by the physiological state of the cell or mycelium. Monitoring of the physiological status is of considerable importance in characterising the individual processes and determining the course of the reaction. Many methods have been described, often adapted for specific applications, and therefore characteristic of a single process. Several practical, and generally applicable, methods will be introduced in this section.

2.1.3.1   **Growth Determination**

The growth of a culture can be determined by measuring several parameters, as follows.

Determination of wet weight. The differing capacities of cells to retain water means that high variability of the results can be expected. The tendency of an organism to take up different amounts of water is on the one hand dependent on the age of the culture and, on the other hand, on the metabolites and cell components, e.g. polysaccharides, polyphosphates, which have a strong influence on the osmotic properties of the media and cells. For these reasons a determination of the wet weight of cells should be made only in exceptional circumstances.

Determination of cell number and size. This method allows very precise deductions to be made about the metabolic status and properties of the organisms, but is somewhat complicated. The problems arising from water retention mentioned above can cause serious mistakes when determining cell size.

The most widely used method is the *determination of dry weight.* By drying the cells to constant weight (this generally takes about 24 h) the errors caused by water retention can be avoided. The filtrate obtained after removing cells or aggregates from the media must be thoroughly washed with water or physiological saline before the drying process is started. If a culture is homogeneous the dry weight can be determined optically by turbidity measurement at 578 nm with the help of a standard curve to determine the dry weight of the culture under investigation. This method is suitable only for very fine, homogeneous cell or spore suspensions, not for those which tend to form cell aggregates or for filamentous organisms. Typical examples of the individual methods are described below.

Determination of Cell Number

For the determination of the cell number of a culture the following (unsterile) equipment and solutions are required: two wide-mouthed or sawn-off glass pipettes (to allow pipetting of large, e.g. plant, cells); automatic pipetting aid; four small test tubes; one 1 ml glass pipette; chromic acid (12.35 g CrO_3, 70 ml ddH_2O). (*Chromium is a heavy metal. All solutions containing chromium must be collected and disposed of according to the safety regulations*); water bath at 60 °C (should contain test tube rack); four long pasteur pipettes and rubber teat; one 10 ml glass pipette; NaCl solution (0.5% (w/v) in ddH_2O); Eppendorf pipette, e.g. for 20 μl; counting chamber, e.g. Fuchs-Rosenthal; microscope.

PROCEDURE

Using the wide-mouthed 2 ml pipette, two 1 ml aliquots are withdrawn from the culture (shaken well to counteract sedimentation) and transferred to two small test tubes. Add 1 ml chromic acid to each probe and place the test tubes in the 60 °C water bath for 20 min. During the incubation the suspensions are constantly mixed by taking up into, and expelling from, a pasteur pipette, thus dispersing any clumps of cells. After the incubation the cells are diluted 1 : 5 to 1 : 20 (depending on the expected number of cells) with 0.5% NaCl. Using the Eppendorf pipette 20 μl of the cell suspension is pipetted onto the counting chamber and the coverslip placed on top. The cells in seven large quadrants (one horizontal row and one vertical row) are counted. After cleaning the counting chamber the procedure is repeated twice. For the calculation of the cell number the volume of the counting chamber and the dilution factor must be taken into account. The volume of a large quadrant in the Fuchs-Rosenthal counting chamber is 0.2 mm^3.

Bacterial Growth

The most convenient method of determining bacterial growth is to measure the extinction of the cell suspension at 578 nm. The dry weight is calculated from the slope of a calibration curve which is obtained as follows.

1   10 centrifuge tubes are marked, dried to constant weight and their weight determined.

2   The appropriate bacterial strain is grown up to the end of the logarithmic phase.

3   The cells are harvested by centrifugation, washed three times with dH_2O, resuspended in dH_2O, and the volume made up to 100 ml (volumetric flask).

4   Using this cell suspension a series of dilutions (10 samples between OD = 0 and OD = 0.4) is made, and the extinction (OD) of each one is determined.

5   From the cell suspension prepared in step 3, pipette 5 ml into each of the centrifuge tubes. After centrifugation the supernatant is carefully removed, preferably by aspiration so that none of the cell material is lost. The cell pellet is dried at 70 °C to constant weight and then each tube is weighed.

6   Weighing of each 5 ml aliquot is carried out after the tubes have been allowed to cool in an exsicator.

7   Finally the calibration curve is drawn and its slope (F) determined.

8   Calculation of dry weight:

$$DW \text{ (mg/ml)} = F \times OD_{578}$$

Determination of Dry Weight

The samples are diluted such that the OD does not exceed 0.3. The measurement of extinction is made at 578 nm. The dry weight is determined with the aid of the calibration curve.

Growth Measurements in Tissue Culture, Fungi, etc.

For dry weight determinations 2–10 ml of the cell or mycelium suspension is filtered through a weighed filter paper which has been dried to constant weight. The cell material remaining on the filter paper is washed carefully with a ten-fold volume of dH_2O, physiological saline or buffer, and the filter paper plus cell material is placed at 80–90 °C until constant weight is reached (up to 24 h) or lyophilised. If the cultures used are homogeneous and tend to form only very small clumps, their growth can be measured by determining the increase in turbidity and plotting a calibration curve. (OD_{578} is measured against medium.)

2.1.3.2   Viability Tests and Staining Techniques

Activity measurements of intracellular reducing enzymes are often used as parameters of cell viability since they reflect the cell's biochemical activity. It should, however, be mentioned that even in instances of reduced cell viability, reductases often retain their activity; therefore exact and critical analysis of the results is essential. For this reason every method must be tailored for each specific case, regarding

- the transport of the precursor into the cells
- the extraction or transport of the product
- pH optimum
- incubation conditions etc.

The following methods are the most frequently used to date.

TTC test. Intact cells reduce 2,3,5-triphenyl tetrazolium chloride (TTC) to water-insoluble triphenylformazane which is red in colour (Zapata *et al.*, 1991). It is possible to monitor the reaction by checking the cells microscopically or by extracting and measuring the red compound. These reactions are oxygen-sensitive. TTC and derivatives thereof are in some instances toxic to cells, and their presence may cause cell damage. The red compound can be extracted from cells by ethanol.

The OD of the red ethanolic solution is in most cases directly proportional to the activity of intracellular reductases. For comparison purposes it is recommended that the OD of the extracted solution be expressed in terms of the dry weight of the cell suspension used. The intensity of the red colour obtained with different cell types increases in the order:

plant/animal cells < fungi < bacteria.

NBT test. Nitro-blue tetrazolium chloride, a derivative of TTC, is less toxic for cells, and the dye compound formed is water soluble. The method is frequently used for plant tissue culture. However, not all cultures give good results. It appears to be particularly well-suited to green plant tissue culture (Thiemann *et al.*, 1989).

TTC Test for the Determination of Reductase Activity

REAGENT

0.1% TTC (w/v) in 0.05 M Na phosphate buffer pH 7.5.

PROCEDURE

1–2 ml of the cell suspension is centrifuged for 5 min at 6000 rpm and the cell pellet is resuspended in 3 ml TTC reagent. The centrifuge tubes are closed tightly (reaction is oxygen sensitive!) and placed in the dark (do not shake!). After this 24 h incubation in the dark at room temperature (22 °C) the samples are centrifuged again and the cell pellet resuspended in about 8 ml ethanol. The suspension is incubated at 60 °C for 15 min. After cooling and centrifugation the supernatant (red) is decanted into a measuring cylinder and the volume made up to 10 ml with ethanol. OD_{485} is measured. Samples should be measured in duplicate.

Other Methods of Determining Cell Viability

These are of a microscopical nature and depend on staining the cells. One can differentiate between viable and non-viable cells because they take up the dyes to a different extent. The reagents used are taken up either by active transport in the viable cell or they diffuse through the cell wall, remaining bound in non-viable cells. Other methods take advantage of intact, active intracellular enzymes, usually hydrolases, which convert compounds taken up either by diffusion or by active transport into coloured or fluorescent derivatives. Some examples are described below.

Fluorescein diacetate (FDA) diffuses through the cell membrane into the cell where it is converted by esterase activity into fluorescein, which accumulates in the cell because it cannot diffuse through the membrane. The presence and activity of the esterases is a measure of cell viability. Procedure is either by microscopy (number of fluorescent cells as determined in the fluorescent microscope divided by the number of cells observed in the light microscope) or by extraction of the fluorescent compound with the appropriate solvent and quantitative determination in a fluorimeter (Keppler *et al.*, 1988; Seitz *et al.*, 1985; Widholm, 1972).

Phenosafranin is not taken up by viable cells; non-viable cells are stained dark red because they take up the dye (Widholm, 1972; Seitz *et al.*, 1985).

15

Methylene blue reacts with reductases to form leuco-methylene blue, which is white. Non-viable cells are stained blue, viable cells produce the white leuco derivative (Schröder, 1991).

Ethidium bromide (*Warning!* This substance is a strong mutagen.) This compound penetrates the cell wall of dead cells and binds to the nucleic acids (fluorescence).

Evans blue: living cells with an intact plasma membrane are impervious to the stain and appear colourless in the microscope, whereas dead cells are stained dark blue (Keppler *et al.*, 1988).

Further reading

BERGMEYER, H.U. (ed.) (1974 ff.) *Methods of enzyme analysis* (Weinheim: VCH).

BRADFORD, M.M. (1976) A rapid and sensitive method for the quantitation of microgram quantities of protein utilizing the principle of protein dye binding. *Anal. Biochem.* **72**, 248–54.

FAWCETT, J.K. and SCOTT, J.E. (1960) A rapid and precise method for the determination of urea. *J. Clin. Pathol.* **13**, 156–9.

GOMORI, G. (1942) A modification of the colorimetric phosphorous determination for use with the photoelectric colorimeter. *J. Lab. Clin. Med.* **27**, 955.

HOATHER, R.C. and RACKHAM, R.F. (1959) Oxidized nitrogen in waters and sewage effluents observed by ultraviolet spectrophotometry. *Analyst* **84**, 548–51.

KEPPLER, L.D., ATKINSON, M.M. and BAKER, C.J. (1988) Plasma membrane alteration during bacteria-induced hypersensitive reaction in tobacco suspension cells as monitored by accumulation of fluorescein. *Physiol. Molec. Plant Pathol.* **32**, 209–19.

KLEBER, H.-P., SCHLEE, D. and SCHÖPP, W. (1990) *Biochemisches Praktikum* (Jena: Gustav Fischer Verlag).

LOWRY, O.H., ROSEBROUGH, N.K., FARR, A.L. and RANDALL, R.J. (1951) Protein measurement with the Folin phenol reagent. *J. Biol. Chem.* **195**, 265–75.

PERNER, B. and SCHMAUDER, H.-P. (1992) Zum Einfluß des bakteriellen Phytotoxins Coronatin auf pflanzliche Zellsuspensionskulturen. *J. Phytopathol.* **135**, 224–32.

ROOS, W. (1989) Kinetic properties, nutrient-dependent regulation and energy coupling of amino-acid transport systems in *Penicillium cyclopium*. *Biochem. Biophys. Acta* **978**, 119–33.

SAWICKI, C.R. and SCARINGELLI, F.P. (1971) Colorimetric determination of nitrate after hydrazine reduction to nitrite. *Microchem. J.* **16**, 657–72.

SCHRÖDER, H. (1991) *Mikrobiologisches Praktikum* (Berlin: Volk und Wissen Verlag).

SEITZ, H.U., SEITZ, U. and ALFERMANN, A.W. (1985) *Pflanzliche Gewebekultur - ein Praktikum* (Stuttgart: Gustav Fischer Verlag).

SÜßMUTH, R., EBERSPÄCHER, J., HAAG, R. and SPRINGER, W. (1987) *Biochemisch–mikrobiologisches Praktikum* (Stuttgart: Georg Thieme Verlag).

SUZUKI, T., YOSHIOKA, T., KATO, Y. and FUJITA, Y. (1987) New estimation method for plant cell viability by determining electron transport activity. *Plant Cell Rep.* **6**, 279–82.

THIEMANN, J., NIESWANDT, A. and BARZ, W. (1989) A microtest system for the serial assay of phytotoxic compounds using photoautotrophic cell suspension cultures of *Chenopodium rubrum*. *Plant Cell Rep.* **8**, 399–402.

WIDHOLM, J.M. (1972) The use of fluorescein diacetate and phenosafranin for determining viability of cultured plant cells. *Stain Technol.* **47**, 189–94.

ZAPATA, J.M., SALMAS, C., CALDERON, A.A., MUNOZ, R. and ROS BARCELO, A. (1991) Reduction of 2,3,5-triphenyltetrazolium chloride by the KCN-insensitive, salicylhydroxamic acid-sensitive alternative respiratory pathway of mitochondria from cultured grapevine cells. *Plant Cell Rep.* **10**, 579–82.

2.2   STRAIN ISOLATION, PROPAGATION AND STORAGE

B. Völksch, Th. Günther and H.-P. Schmauder

2.2.1   *Selection and Isolation of Microorganisms (using Bacteria as an Example)*

The isolation of microorganisms from their natural environment is a prerequisite for dealing with many biotechnologically relevant problems. The aim of the isolation can be either a *pure culture*, i.e. a microbial strain originating from a single cell, or a *mixed culture* consisting of species of the natural population.

In the wild there are few habitats where one can find microorganisms of a single species in a concentration sufficiently high to be regarded as a 'pure culture'. Therefore the isolation of microorganisms usually begins with a 'selection culture' leading to an increase in the number of bacteria of the desired species or genus above that of the other species making up the original mixed culture. By the use of so-called selective media or by choosing specific culture conditions (temperature, O_2 concentration, pH) it is possible to optimise the growth conditions for one or at most a few species so that these can outgrow other, for the purpose in question, undesirable microorganisms. An enrichment for particular organisms can be achieved by preventing growth (e.g. by the addition of antibiotics) or selective killing (e.g. pasteurisation) of the undesired microorganism.

Enrichment cultures are generally prepared in liquid medium. The sample material (sewage slurry, soil, waste water) is added to the selective culture medium and incubated before an aliquot of the grown culture is transferred to fresh medium and incubated further. This procedure is repeated and the culture examined

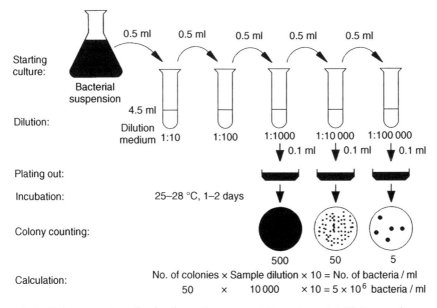

Figure 2.1   Colony count method: schematic representation of a serial dilution and calculation of the number of viable cells.

microscopically until the culture contains microorganisms of only one, or at most only a few types.

The subsequent *purification* results in a separation of the individual bacterial cells. In the case of aerobic bacteria this is usually achieved by plating out on solid media (Fig. 2.1) or by streaking out for single colonies using a platinum needle. The single colonies obtained on incubation have to be purified until only colonies of uniform shape, size, colour and consistency are obtained. The pure culture is also tested by microscopy of stained preparations. The isolation of microorganisms (aerobic and anaerobic bacteria, fungi and yeast) can be performed by various methods, which are not all described here in detail, but can be found in the literature.

Isolation of Bacteria Capable of Degrading Polycyclic Aromatic Hydrocarbons (PAH) from Oil-contaminated Earth

Earth contaminated with organic waste material is a naturally occurring source of microorganisms with degradative properties. Bacteria capable of degrading polycyclic aromatic hydrocarbons (PAH) can be isolated from such locations. Earth samples are transferred to liquid medium which contains, as sole energy and carbon source, the substance to be degraded. Successive transfer of aliquots of the growing culture into fresh medium selects for those bacteria capable of using, and therefore degrading, the carbon source in the medium. The poor water solubility of PAHs is exploited for the isolation of degradative strains on plates.

The aim of the experiment is to enrich for, and isolate, a bacterial strain capable of degrading the environmental pollutant, phenanthrene. It must be demonstrated that the strains selected can grow on phenanthrene.

Principle

Samples (removed from just below the surface) of earth contaminated with oil (e.g. near filling stations, car parks or railway embankments) are transferred to liquid medium containing the three-ringed PAH phenanthrene (toxic!) as sole carbon and energy source. Once growth of the microorganisms has been established, phenanthrene-degrading bacteria are selected by repeated passages through the selective liquid medium. Suitably diluted aliquots of the bacteria are spread onto plates which are then sprayed with a solution of phenanthrene, and those colonies surrounded by a clear 'halo' are picked. These degradative strains are grown in the presence of phenanthrene as the sole carbon source. The growth of the strain and the reduction in the amount of the substrate are determined.

Media

Nutrient agar: made up according to the manufacturer's instructions. *Liquid enrichment medium solution*: KH_2PO_4 1.0 g, $Na_2HPO_4 \times 12\ H_2O$ 1.25 g, $(NH_4)_2SO_4$ 1.0 g, $MgSO_4 \times 7\ H_2O$ 0.5 g, $CaCl_2 \times 6\ H_2O$ 0.05 g, $FeSO_4 \times 7\ H_2O$ 0.005 g, phenanthrene 0.05 g, dH_2O ad 1000 ml, pH 6.8–7.0. Dissolve the salts separately in dH_2O, combine the solutions and make up to 1 l; adjust pH, dispense 20 ml aliquots into 100 ml Erlenmeyer flasks and autoclave. Dissolve 50 mg phenanthrene in 5 ml methanol and add 0.1 ml of this solution to 20 ml sterile liquid enrichment solution.

Phenanthrene forms a finely divided precipitate (solubility of phenanthrene in water is approximately 1 mg/l). *Inoculation medium*: as liquid enrichment medium but with 3 g glucose/l instead of phenanthrene.

Equipment

100 ml Erlenmeyer flasks containing 20 ml liquid enrichment medium; screw-capped or glass-stoppered test tubes containing 5 ml liquid enrichment medium or inoculation medium; agar slants; nutrient agar plates; sterile tubes containing 4.5 ml 0.9% NaCl; sterile centrifuge tubes; microscope; counting chamber for bacteria; sterile pipettes; sterile spreader; wash bottle; 1% phenanthrene solution in ether; 0.1 M potassium phosphate buffer pH 7.5; HPLC.

Procedure

ENRICHMENT OF PHENANTHRENE-DEGRADING BACTERIA

1   Inoculate each of five Erlenmeyer flasks containing liquid enrichment medium with a small quantity (just the amount which can be held on the end of a spatula) of soil. The soil sample should have a distinct 'oily' smell. The flasks are incubated, with shaking, at 28 °C until the medium becomes turbid as a result of bacterial growth (2–7 days).

2   Carry out regular microscopical monitoring of the culture.

3   Transfer 1 ml of the culture into fresh medium and examine the culture in the microscope as soon as it becomes turbid. Inoculate 1 ml of the culture into fresh medium.

4   Repeat the passage of the enrichment culture twice.

ISOLATION OF PHENANTHRENE-DEGRADING BACTERIA

5   Prepare serial dilutions (up to 10^{-5}) of the culture.

6   Add 0.5 ml culture to 4.5 ml 0.9% NaCl and mix well (10^{-1} dilution).

7   Using the 10^{-1} dilution prepare a 10^{-2} dilution in the same way and repeat for further tenfold dilutions.

8   Plate 0.1 ml of the 10^{-3} to 10^{-5} dilutions on nutrient agar plates. The plates are incubated at 28 °C until colonies are visible (approx. 2–3 days).

9   Plates with well-separated colonies are sprayed lightly with the ether solution of phenanthrene, so that as soon as the ether evaporates a visible layer of water-insoluble phenanthrene remains.

10 The plates are incubated at 28 °C in a humidity chamber for several days. The plates are checked daily; colonies made up of bacteria capable of degrading phenanthrene are surrounded by a clear 'halo'; phenanthrene is taken up by the cells and degraded, causing the agar in the immediate vicinity of the colony to appear clear.

11 Material from colonies surrounded by 'haloes' is streaked for single colonies onto nutrient agar, and the plates are incubated at 28 °C. If necessary the streaking procedure is repeated.

12 Prepare stock cultures from colonies of different appearance (colour, consistency, morphology) by streaking onto agar slants.

13 Identify the isolates (see Subsection 2.2.1.3).

GROWTH OF CULTURES ON PHENANTHRENE

1 To prepare a starting culture, inoculate cell material from an agar slope (2–7 days old) into inoculation medium and incubate at 28 °C with shaking for 24 h.

2 Harvest cells by centrifugation (10 min, 6000 rpm) and resuspend the cell pellet in 5 ml potassium phosphate buffer. Inoculate 10 test tubes, each containing 5 ml enrichment medium, with 0.1 ml of the cell suspension. (The number of cells in the inoculum should lie between 10^6 and 10^7.)

3 Two test tubes serve as controls and are examined immediately as described below, and the remaining eight are incubated at 28 °C. Every 2 days two test tubes are removed and examined as follows:

 (i) *Determination of cell number using a cell counting chamber.* After the addition of one drop of Tween 20 each culture is 'vortexed' to disperse any cell aggregates which have formed and to disrupt any cell–substrate interactions. One drop of this cell suspension is placed in the counting chamber and the cell number determined.

 (ii) *Determination of phenanthrene concentration.* 5 ml methanol is added to each test tube to bring the phenanthrene into solution. After mixing thoroughly the samples are allowed to stand for 10 min before centrifuging at 6000 rpm for 10 min. Carefully remove 1–2 ml of the supernatant for the determination of phenanthrene by HPLC analysis. All measurements are done in duplicate using a standard solution of phenanthrene (50 mg/l phenanthrene dissolved in methanol).

Separation and detection conditions: RP-18 column, 125 mm; acetonitrile : water = 80 : 20 (v/v); flow rate: 1.5 ml/min, isocratic; UV detector at 254 nm.

Interpretation

Determine cell number and phenanthrene concentration in media. Bacterial growth accompanied by a decrease in phenanthrene concentration is evidence that phenanthrene is being used as the sole energy and carbon source. These observations must be supported by further experiments.

Isolation and Characterisation of Bacteria from Leaf Tissue

Knowledge of the ecology of microbes, i.e. their interaction with their natural environment, is a prerequisite if microbial processes are to be fully understood and exploited. This potential source of natural products is a virtually unexplored reservoir of biologically active compounds. The aim is to isolate novel and/or known bacteria with hitherto undiscovered biochemical capabilities (e.g. with ability to synthesise active and valuable compounds). Particularly important is the isolation of specialised bacteria from their natural environment. The model system chosen for the following experiments is leaf rot.

Isolation of Bacteria from the 'Micro-environment' Leaf Rot

PRINCIPLE OF THE EXPERIMENT

Conclusions about microorganisms cannot always be drawn from the damage they cause. Apparently identical or similar lesions may have arisen for totally different reasons, not necessarily biological. Therefore not only must the damage be recognised, but also the microorganisms responsible must be isolated and identified.

MATERIAL AND MEDIA

Crystal violet agar: crystal violet 2.0 mg, sucrose 50.0 g, nutrient agar 50.0 g, dH$_2$O ad 1000 ml. *Physiological saline*: NaCl 0.85%. *Broth-glycerol-agar (BGA)*: peptone 5.0 g, yeast extract 1.0 g, glycerine 20.0 g, Liebig's meat extract 3.0 g, agar 18.0 g, dH$_2$O ad 1000 ml, pH 7.0.

PROCEDURE

1 5 mm × 5 mm fragments are excised from an infected leaf; if possible the tissue samples should be taken from areas where infected and healthy tissue are to be found side by side.

2 The tissue samples are disinfected by rinsing in ethanol. They are washed in sterile dH$_2$O before homogenisation. 3 ml dH$_2$O is added to the homogenate.

3 0.1 ml of the suspension obtained in step 2 is plated onto crystal violet agar.

4 The plates are incubated at 28 °C for 24–48 h.

5 The bacteria which have grown on the crystal violet agar are streaked out for single colonies. Single colonies are then transferred to BGA.

6 Agar slopes prepared from these isolates will serve as starting material for further cultivation and identification of the bacteria.

COMMENTARY

At a concentration of 0.0002% crystal violet inhibits the growth of Gram-positive bacteria. Crystal violet agar therefore is an enrichment medium for Gram-negative bacteria commonly found in leaf tissue or associated with plants.

Proof that the Isolated Bacteria Produce Biologically Active Substances

PRINCIPLE OF THE TEST

Preliminary evidence for the production of biologically active metabolites by the individual microorganisms is obtained from 'biotests'. Biotests are an essential tool for the discovery of biologically active substances and the characterisation of their structure and function. Biotests must be specific, sensitive, simple and reproducible. The active substance is defined 'biologically' prior to explaining its chemical structure. The choice of tests is dependent on the type of investigation, the nature of the active substance and the 'indicator organism', and no 'hard and fast' rule can be followed.

The test described below is based on the diffusion of an active substance through agar and the consequent formation of a concentration gradient. This influences the growth of the indicator strain (e.g. *E. coli* to test for bacteriocidal effect; *Chlorella pyrenoidosa* to test for the effect on algae and the photosynthetic apparatus) as shown by the formation of a ring of growth inhibition. The diameter of this ring is a measure of the concentration of the active substance. There is a linear relationship between the diameter of the circle of growth inhibition and the $\log_{10}$ of the concentration of the active substance.

Therefore the effective concentration can be determined with the aid of a standard curve. The test must be standardised, and this demands the utmost precision to give reproducible results.

STRAINS AND MEDIA

Strains. Escherichia coli N 100, Chlorella pyrenoidosa. Media. Minimal medium: KH_2PO_4 1.0 g, K_2HPO_4 1.0 g, NaCl 1.0 g, Na citrate 1.0 g, $(NH_4)_2SO_4$ 4.0 g, glucose 2.0 g, $MgSO_4 \times 7\ H_2O$ 0.7 g (to be added after the other salts have been dissolved), dH_2O ad 1000 ml; agar if necessary, 20.0 g. *Medium 5b*: glucose 8.8 g (sterilise separately), KH_2PO_4 2.6 g, Na_2HPO_4 5.5 g, NH_4Cl 2.5 g, Na_2SO_4 1.0 g, $MgSO_4 \times 7\ H_2O$ 0.1 g, KCl 0.1 g, $FeSO_4 \times 7\ H_2O$ 0.01 g, $MnSO_4 \times 4\ H_2O$ 0.01 g, dH_2O ad 1000 ml. *Kandler-Ernst medium*: glucose 10.0 g, KNO_3 0.4 g, $Ca(NO_3)_2 \times 4\ H_2O$ 0.1 g, $MgSO_4 \times 7\ H_2O$ 0.1 g, KCl 0.1 g, K_2HPO_4 0.1 g, Fe-EDTA 7.0 mg (!), agar 20.0 g, dH_2O ad 1000 ml. This medium without the addition of glucose can be used for the propagation of algae.

PROCEDURE

A: *Cultivation of the strains to be tested*. Cultivation is carried out at 18 °C and 28 °C in 100 ml straight-sided flasks containing 20 ml medium 5b. 5 ml sterile dH_2O is pipetted into each of the agar slopes of the isolates prepared in step 6 above to resuspend the cells. 1.0 ml of the cell suspension serves as the inoculum. Duplicate cultures are incubated with shaking (250 rpm) at 18 °C and 28 °C for 24 h. 1.0 ml of these cultures is inoculated into 20 ml medium 5b and incubated as described above for 48 h. The growth of the main culture ($X = X_{48} - X_0$) is determined by measuring the extinction at 578 nm. The cells are harvested by centrifugation (20 min, 6000 rpm). Heat-treated (5 min, 120 °C, pressure cooker) and untreated supernatants are used for the biotest.

B: *E. coli biotest*. The cells of a 24-h-old nutrient agar culture of *E. coli* are transferred to 100 ml minimal medium contained in a 500 ml flat-bottomed round flask. This culture is incubated at 28 °C with shaking (250 rpm) for 16–24 h. Immediately prior to the start of the test the OD_{578} of this culture is adjusted to 0.6 by the addition of sterile minimal medium. Test plates (diam. 130 mm) are prepared by combining (for each plate) 40 ml 2% minimal agar medium (auto-claved and cooled to 45 °C) with 2 ml of the diluted *E. coli* culture. After pouring, the plates are placed in the refrigerator for 30–60 min. Circular wells (diam. 8 mm) are made in the agar. 50 μl of the solutions to be tested are pipetted into the wells. The plates are incubated at 37 °C for 24 h before measuring the ring of growth inhibition.

C: *Chlorella biotest*. The test plates (130 mm diam.) are two-layered; the lower layer consists of 40 ml *Kandler–Ernst medium*. Once the agar has solidified, the plates are dried at 37 °C for 1 h and stored at room temperature for approximately 16 h. The upper layer is made up of 4 ml 1.5% water agar (autoclaved and cooled to 45 °C) mixed with 4 ml of a suspension of algae prepared from a 14-day-old agar culture. The algal suspension is obtained by washing the cells off the agar slope with dH$_2$O and adjusting the OD$_{440}$ to 0.5.

Circular wells, 8 mm in diameter, are made in the agar. 50 μl of the suspensions to be tested are pipetted into the wells and the plates are incubated for 2–4 days at 28 °C under constant illumination. The extent of the growth inhibition can be quantified from the size of the zones surrounding the wells.

COMMENTARY

The effect of the test supernatants on the indicator organisms is summarised as follows:

• bacteriocidal effect

• anti-algal effect

• damage to the photosynthetic apparatus.

Furthermore, preliminary information on the possible chemical structure of the active substance(s) can be obtained.

Taxonomic Classification of the Isolated Bacterial Strains

The classification of strains is not only important for the diagnosis of pathogens of man, animals and plants, but also for biotechnology and ecology. Biotechnology needs 'named' strains with defined characteristics. A knowledge of the properties of a production strain is essential for quality control and if the strain is to be 'improved'. A correct identification of the production strain allows the user to benefit from the relevant ecological, medical, genetical and biochemical data available. The importance of systematics for the biotechnologist cannot be overemphasised.

Principle

Specific characteristics of strains, allowing their identification as members of a family or at least a group, as defined in *Bergey's manual of systematic bacteriology*, Vols 1–4 (1984–1989) are determined in a series of basic tests.

Strains, Media and Equipment

Strains. *Escherichia coli* (e.g. DSM 498), *Bacillus subtilis* (e.g. DSM 10), *Pseudomonas fluorescens* (e.g. DSM 50090), *Streptococcus* spp. (e.g. DSM 20067).

Media. *Nutrient agar* (prepared according to the manufacturer's instructions); sterile paraffin; KOH solution (3%); dimethyl-*p*-phenylenediamine-HCl (1%); H$_2$O$_2$ (3%). *Hugh-Leifson agar*: peptone 0.2%, NaCl 0.5%, K$_2$HPO$_4$ 0.03%, agar 0.3%, Bromothymol blue (1.0% stock solution) 0.003%, glucose 1.0%, pH 7.1. The glucose solution (10.0% in dH$_2$O) should be filter-sterilised and added to the molten agar after cooling.

A: Gram typing by means of the KOH rapid test according to Gregersen. This test depends on the fact that, under the test conditions, Gram-positive cells burst and release their DNA. Resuspend a small amount of cell material in one drop of 3% KOH solution and mix thoroughly. After 5–10 s Gram-negative bacteria (e.g. *E. coli*) give a positive reaction: slimy threads 1–2 cm long can be drawn with the inoculating needle. If no slimy threads can be drawn with the inoculating needle the strain is considered to be Gram-positive (e.g. *B. subtilis*).

B: Determination of cell morphology. A live preparation of the strain is made and examined, at a magnification of 100 × under oil immersion, in a phase contrast microscope. Shape, size (measured with a calibrated objective micrometer), formation of aggregates, spore formation (shape and position within the cell) and motility are noted.

C: Determination of colony morphology. The strains are streaked out for single colonies on nutrient agar and incubated at 28 °C for 48–72 h. Colour, size, shape, surface of the colony and secretion of pigments should be noted.

D: Catalase test. The enzyme catalase is found in strictly aerobic and many facultative anaerobic bacteria. The lack of superoxide dismutase and catalase is responsible for the oxygen sensitivity of strict anaerobes. The enzyme catalyses the breakdown of H_2O_2 into oxygen and water.

1   A small amount of the strain to be tested is placed on a microscope slide; on top of this is placed one drop of 3% H_2O_2.

2   The formation of bubbles (O_2) is a sign that the strain is capable of producing catalase.

3   Catalase-positive: *Pseudomonas fluorescens*; catalase-negative: *Streptococcus* spp.

E: O/F test with glucose. In this test developed by Hugh and Leifson in 1953 the capacity of a bacterial strain to use a specific sugar in a complex medium either oxidatively or fermentatively is examined. It is possible to test simultaneously if the pH is reduced (the indicator turns yellow) or if a gas (bubble formation) is produced. This test allows one to check the use of different mono- and disaccharides or sugar alcohols by bacteria.

1   Two test tubes, each containing Hugh-Leifson medium are inoculated by inserting an inoculating needle into the soft agar.

2   One tube is rendered airtight by overlaying with 2 ml of sterile paraffin.

3   Both cultures are incubated at 28 °C.

4   The test tubes are examined after 2 and 6 days incubation.

5   A yellow coloration indicates that the sugar is metabolised under the relevant conditions of incubation.

6   Oxidative: *Pseudomonas fluorescens*; fermentative: *Escherichia coli*.

F: Oxidase test. The test described by Kovacs (1956) depends on the enzyme cytochrome oxidase. Cytochrome oxidase is the last enzyme in the respiratory chain and is located in the membrane of several bacteria. The reagent dimethyl-*p*-phenylenediamine-HCl acts as an electron donor for cytochrome-*c*-oxidase, which then transfers the electrons to molecular oxygen. The product of the reaction is blue.

1 The bacteria to be tested are streaked onto filter paper soaked in a 1% solution of dimethyl-*p*-phenylenediamine-HCl.

2 The test should be performed with cultures which are older than 24 h.

3 A blue/purple coloration appearing within 10 s is deemed positive; a reaction time of up to 60 s is still regarded as positive.

4 Oxidase-positive: *Pseudomonas fluorescens*; oxidase-negative: *Escherichia coli*.

Commentary

The results of the individual tests are presented in a tabular form, so that a quasi-taxonomic ordering is possible. The tests must always be performed with reference strains as a control to ensure that the reagents are correct. Using data from the literature it is possible to classify a strain down to the race.

Further reading

ALEF, K. (1991) *Methodenhandbuch Bodenmikrobiologie* (Landsberg/Lech: Ecomed).

BOARD, R.G. and LOVELOCK, D.W. (1975) *Some methods for microbiological assay* (London: Academic Press).

BRADBURY, J.F. (1970) Isolation and preliminary study of bacteria from plant tissue. *Rev. Plant Pathol.* **49**, 213–18.

COOK, A.M., GROSSENBACHER, H. and HÜTTER, R. (1983) Isolation and cultivation of microbes with biodegradative potential. *Experientia* **39**, 1191–8.

DREWS, G. (1983) *Mikrobiologisches Praktikum* (Berlin: Springer).

DURBIN, R.B. (1981) *Toxins in plant disease* (New York: Academic Press).

GASSON, M.J. (1980) Indicator technique for antimetabolic toxin production by phytopathogenic species of *Pseudomonas*. *Appl. Environ. Microbiol.* **39**, 25–9.

GREGERSEN, T. (1978) Rapid method for distinction of gram-negative from gram-positive bacteria. *Europ. J. Appl. Microbiol. Biotechnol.* **5**, 123–7.

HUGH, R. and LEIFSON, E. (1953) The taxonomic significance of fermentative versus oxidative metabolism of carbohydrates by various gram-negative bacteria. *J. Bacteriol.* **66**, 24–6.

KIYOHARA, H., NAGOA, K. and YANA, K. (1982) Rapid screen for bacteria degrading water-insoluble, solid hydrocarbons on agar plates. *Appl. Environ. Microbiol.* **43**, 454–7.

KOVACS, N. (1956) Identification of *Pseudomonas pyocyanes* by the oxidative reaction. *Nature* **178**, 703.

KREISEL, H. and SCHAUER, F. (1987) *Methoden des mykologischen Laboratoriums* (Jena: Gustav Fischer Verlag).

KRIEG, N.R. and HOLT, J.G. (eds) (1984) *Bergey's manual of systematic bacteriology*, Vol. 1 (Baltimore: Williams and Wilkins).

SNEATH, P.H.A., MAIR, N.S., SHARPE, M.E. and HOLT, J.G. (eds) (1986) *Bergey's manual of systematic bacteriology*, Vol. 2 (Baltimore: Williams and Wilkins).

STANLEY, J.T., BRYANT, M.P. PFENNIG, N. and HOLT, J.G. (eds) (1989) *Bergey's manual of systematic bacteriology*, Vol. 3 (Baltimore: Williams and Wilkins).

WILLIAMS, S.T., SHARPE, M.E. and HOLT, J.G. (eds) (1989) *Bergey's manual of systematic bacteriology*, Vol. 4 (Baltimore: Williams and Wilkins).

2.2.2 *Long- and Short-term Storage of Microorganisms*

Microorganisms are maintained as stock cultures for short-term storage or conservation cultures for long-term storage. In both forms of storage the cultures must remain viable, retain their strain characteristics and be free from contamination. There is no single method of storage applicable to all types of microorganisms. The most appropriate method has to be found by trial and error if none of the methods described in the literature proves satisfactory.

Stock Cultures

In stock cultures the cells are maintained in a state of reduced metabolic activity, from which they can be readily revived. The techniques involved are simple, but the cultures must be transferred to fresh medium at regular intervals. Depending on the organism, this time interval can range from days to a year. The possibility that the strain will mutate under such storage conditions is relatively high.

Methods for Maintaining Stock Cultures

- agar slopes
- agar stabs
- glycerol cultures held at $-20\,°C$

Conservation Cultures

The strains can be kept in these cultures over a number of years. Under the conditions chosen there is little or no chance of the strains undergoing mutational changes. The methods used for long-term storage tend to be complicated, and revival of some microorganisms from this latent state can prove difficult.

Methods for Maintaining Conservation Cultures

1 at ambient temperature
   (i)   drying from the liquid state
   (ii) lyophilisation
   (iii) silica gel, porcelain or glass bead cultures
   (iv) conservation in gelatine
2 at low temperature
   (i)   storage at $-196\,°C$ in liquid nitrogen
   (ii) storage at $-80\,°C$ in glycerine
   (iii) storage at $-80\,°C$ with glass beads
3 storage in paraffin oil
4 storage in sterile soil.

Storage of Strains on Agar Slopes

The maintenance of strains by restreaking on fresh medium at regular intervals over a period of years is the 'classical' method of storing a culture and is particularly suitable for many bacterial strains.

Procedure

1   Prepare sterile agar slopes.
2   Using a flamed and cooled ('jabbed' into agar) inoculation needle, spread the strain to be stored on the surface of the agar.
3   Incubate the agar slope at the appropriate temperature until growth is visible (1–2 days).
4   Wrap the agar slope in aluminium foil (to prevent drying out) and store at 4 °C.

Further Experiment

Test the viability of the strain after several weeks by restreaking onto a fresh agar slope.

Conservation of Bacteria by Lyophilisation

Lyophilisation is the method of choice for maintaining bacterial strains with the least possibility of their properties being altered. The bacterial suspension is frozen before being subjected to drying *in vacuo*.

Procedure

1   Prepare the lyophilisation medium by centrifuging milk at 4000 rpm for 20 min. Decant the supernatant, and withdraw and sterilise twice (15 min, 120 °C) 10 ml of whey. Cover two glass vials per strain with aluminium foil and sterilise.
2   Inoculate the strains onto agar slopes, and incubate for 24 h at 28 °C.
3   Resuspend the cells in the sterilised whey. Using an Eppendorf pipette transfer 0.2 ml of the cell suspension into each vial.
4   Place vials at −20 °C.
5   Lyophilise samples and seal the vials using a Bunsen burner.
6   Store vials at 4 °C.

Experimental control

Determine the number of viable cells before and after several weeks in the lyophilised state.

Conservation of Bacteria by Freezing in the Presence of Glycerine

To date, conservation below freezing point is the best way of maintaining the viability, physiological properties, biochemical activity and genetic stability of microorganisms over an extended period of time. At temperatures around −80 °C water molecules are still in the liquid state, so that the cells have to be protected by the addition of suitable compounds to the freezing medium.

Procedure

1   Sterilise small screw-capped glass bottles containing 0.5 ml 86% glycerine.
2   Grow the strain to be conserved in rich liquid medium to the late logarithmic phase.
3   Pipette 2.5 ml of the culture into the prepared glycerine vials, close tightly and mix thoroughly.
4   Store at −80 °C.

Commentary

Determine the viable cell count immediately before storage in glycerine and again a few weeks later. (Dilution!)

Conservation of Fungal Strains by Freezing

Fungal strains which produce secondary metabolites have a marked tendency to 'degenerate'. After several 'passages' on solid or in liquid medium, valuable physiological properties may be lost. For this reason such strains have to be conserved with particular care and attention. Often the problem can be solved by conservation of the strains at low temperature. The method of conservation decided upon depends on whether or not the strain produces spores. Generally the capacity to form spores and the ability to form secondary metabolites are mutually exclusive, and in these instances the best method is to freeze vegetative mycelia.

When the spores have formed these are separated from the mycelia, washed with physiological saline and resuspended in the appropriate storage solution. If there is no spore formation, growing vegetative mycelia should be used as the starting material. As much liquid as possible should be removed by filtration, the mycelia washed and, if possible, homogenised gently in the storage solution. *Storage solutions* are a mixture of carbohydrates and polymers, e.g. albumins, milk proteins and, in some instances, dimethyl sulphoxide. Good results have been achieved using mixtures of 60% sucrose solution containing skimmed milk, in the proportion 2 : 1. The following are important.

*   The solutions should be autoclaved separately or otherwise sterilised.
*   The skimmed milk products used should be free of temperature-resistant spore-forming bacteria. (Check the sterility on nutrient agar)
*   The solutions should be sterilised immediately before use.

Contamination with heat-stable, spore-forming bacteria can be reduced to a minimum by using skimmed milk powder rather than the fresh product. Caramelisation of the sucrose solution after sterilisation generally has no adverse effect. The cells (or homogenate) or spores are resuspended at an appropriate concentration in the freshly made-up storage solution, and the suspension is dispensed into thick-walled, round-bottomed vials (the vials which generally have a volume of 15 ml should contain no more than 5 ml of cell suspension), sealed and placed in the freezer. Storage at temperatures between −35 °C and −80 °C is recommended. The best results are obtained with storage above or in liquid nitrogen. A temperature of $-15\,°C$ should be reached as soon as possible.

Safety! During storage in liquid nitrogen over long periods of time the cooling substance can diffuse into the vials via hairline cracks. Expansion upon thawing can cause the vial to explode! Storage cultures should be brought to room temperature as rapidly as possible.

As soon as the vial has been opened the cell material is transferred to a suitable rich medium.

Stability of the stocks is at least 3–5 years. For this reason conservation over liquid nitrogen is the method of choice for strain preservation in culture collections.

Tests

- determination of the survival rate
- determination of the capacity to form secondary metabolites
- selection of 'over-producers'. (It should be remembered that 'degeneration', which can also occur at low temperatures or in lyophilysates, can lead to a reduced production of secondary metabolites.)

Conservation of Fungi by Lyophilisation

The preparations for lyophilisation are the same as those described for freezing, and the same solutions can be used. The cell suspension is dispensed into the vials which are then covered with loose sterile caps. The vials are frozen at $-30\,°C$ before placing in the freeze-drying apparatus. Without using the heating coil the preparations are dried *in vacuo* over a period of 2–3 days, by which time the temperature inside the vessel has reached room temperature. Prior to sealing, the vials are flooded with nitrogen. The lyophilised cultures have a residual moisture of approximately 4–5% and may be stored either in the refrigerator or at room temperature.

The lyophilised cultures can be reconstituted by transferring some of the lyophilysate to an appropriate medium.

Lyophilised cultures are generally stable over a period of 2–3 years. Alterations in genotype have been reported, but their cause is unknown (Sharp, 1984).

Tests

As described for strains stored at low temperature.

2.2.3 *Fungal and Plant Protoplasts*

Many experiments necessary for explaining biological, biochemical and biotechnological reactions are hampered by the presence of the cell wall in fungi and plants. To overcome this problem, several methods have been developed for the removal of the cell wall to create protoplasts in which the cell membrane is the only barrier to the cytoplasm. As a reaction to their increased sensitivity to external influences, protoplasts have a tendency to regenerate their cell walls as quickly as possible. However, during the period when new cell wall is being formed, there is ample opportunity to investigate the cell's metabolism, to perform mutagenesis etc. using the protoplasts. It should be noted that the newly synthesised cell wall can differ structurally and in its composition from the 'established' version. A schematic representation of the method for producing protoplasts and ways of using them is shown in Fig. 2.2. Using cellular material (plant tissue culture, plant tissue, microorganisms) or filamentous fungi, the cell wall is digested by the action of enzymes capable of breaking down its constituents. The enzyme chosen depends on the organism being used and the composition of its cell wall. It is often necessary to use mixtures of enzymes since cell walls consist of several polymers. Generally, naturally occurring mixtures are used, such as snail gut enzyme, also known as 'helicase', pectinases, cellulases or preparations of mixed microbial exoenzymes specific for cell wall lysis. The exoenzymes can be isolated from cell walls prepared from cultures induced for enzyme production. Such lytic enzymes can be isolated from the spent medium of cultures of various microorganisms (e.g. white rot fungus, brown rot fungus, *Penicillium* spp. and diverse bacterial spp.). Preparation of bacterial protoplasts requires the removal of the outer murein layer by means of lysozyme.

The preparation of protoplasts enzymatically means that certain requirements must be fulfilled, as follows.

- All working solutions must be hypertonic to maintain osmotic stability. This is achieved by the addition of salts, sugars or sugar alcohols (e.g. 0.5 to 0.84 M mannitol or sorbitol).

- All manipulations must be carried out as carefully as possible to reduce mechanical stress on the protoplasts, especially centrifugation.

- The protoplasts should be carefully separated from cell debris, spores and mycelial fragments by filtering through cheesecloth or glass wool.

- An aseptic technique is necessary.

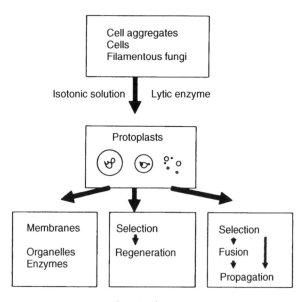

Figure 2.2 Schematic representation of protoplast preparation.

Examples for the isolation of fungal protoplasts are given in Table 2.1. Purified protoplast preparations, which can also be obtained by low-speed centrifugation in isotonic solutions, have multiple uses in research and development. For instance, it is possible by altering the osmotic strength of the solution to lyse the protoplasts gently and release fragile cellular organelles such as vacuoles and membranes or even enzymes. A further possibility is the selection of specific properties (cell contents, biosyntheses) before regeneration of the cell wall sets in. 'High producers' can also be exploited by this technique. Fusion of protoplasts from different sources is also possible. Using this technique, over-producing strains, strains with new properties and intergeneric forms are attainable. When performing fusion experiments it is advisable to have characteristic markers for each partner (e.g. auxotrophy, resistance to antibiotics). Without such selection markers the selection and characterisation of the fusion products are virtually impossible. Fused strains can be propagated as soon as the cell wall has regenerated itself. The following methods are used for cell fusion.

- Electrofusion by the reversible breakdown of cell membranes by the application of an electric field in the range of 1–2 V per ns to 1–2 V per ms. In this method the conductivity and therefore the permeability of the cell membrane is altered, and this allows the protoplasts to fuse with each other. It is also possible that DNA fragments, organelles and large macromolecules for which the cell membrane is normally impermeable can enter into protoplasts.
- Polyethylene glycol is used to facilitate protoplast fusion.

Further information can be found in: Amory and Rouxhet (1989), Aoyaki *et al.* (1993), Bengochea and Dodds (1986), Gleba and Sytnik (1984), Pilet (1985), Stahl *et al.* (1977), Weber (1993).

Table 2.1 Examples of protoplast preparation from various organisms.

Organism	*Claviceps fusiformis, C. purpurea*	*Saccharomyces cerevisiae, S. carlsbergensis*	*Catharanthus roseus, Nicotiana tabacum*
Enzyme used	Helicase (from *Helix pomatia*)	Zymolyase 20T (Japan)	Macerozyme (Japan) Cellulase Onozuka (Japan)
Method of preparation	Shake, 28–30 °C for 3–4 h. Filter through cotton wool, gradient centrifugation (10 min, 1000 g) 5–7 × 10^8 protoplasts per 100 mg wet weight	Stationary phase cells, incubate with 2.5 mM mercaptoethanol, 30 °C, ca. 90 min. Lysis at 35 °C for no longer than 150 min; amount of enzyme used: 15 mg/l	Collect cells by filtration and plasmolyse (30 min, 0.6 M mannitol), 1–2 h incubation with the enzyme mixture
Hypertonic solutions	0.7 M KCl or 1 M MgSO$_4$	1 M sorbitol, 12.5 mM MOPS buffer, 0.05 mM CaCl$_2$	0.6 M Mannitol, 5 mM CaCl$_2$, 1 mM MgCl$_2$, 1 mM MES
Stability	Regeneration starts after 2–3 h	Stability of *S. cerevisiae* < *S. carlsbergensis*	
References	Stahl *et al.* (1977)	Amory and Rouxhet (1989)	Aoyaki *et al.* (1993)

References

AMORY, D.E. and ROUXHET, P.G. (1989) Kinetics of the transformation of *Saccharomyces cerevisiae* and *carlsbergensis* into protoplasts. *Louvain Brew. Lett.*, **3–4**, 16–21.

AOYAKI, H., JITSUFUCHI, T. and TANAKA, H. (1993) Development of an optimal method for monitoring protoplast formation from cultured plant cells. *J. Ferment. Bioeng.* **75**, 201–6.

BENGOCHEA, T. and DODDS, J.H. (1986) *Plant protoplasts – a biotechnological tool for plant improvement* (London: Chapman and Hall).

DIETTRICH, B. and LUCKNER, M. (1981) Die Tieftemperturkonservierung pflanzlicher und tierischer Zellen. *Biol. Rdsch.* **19**, 269–84.

GLEBA, Y.Y. and SYTNIK, K.M. (1984) *Protoplast fusion – genetic engineering in higher plants* (Berlin: Springer Verlag).

HECKLY, R.J. (1985) Principles of preserving bacteria by freeze-drying. *Dev. Ind. Microbiol.* **26**, 379–95.

KIRSOP, B.E. and DOYLE, A. (1991) *Maintenance of microorganisms and cultured cells* (London: Academic Press).

KIRSOP, B.E. and SNELL, J.J.S. (1984) *Maintenance of microorganisms – a manual of laboratory methods* (London: Academic Press).

LUDLUM, H.A., NWACHUKWU, B., NOBLE, W.C., SWAN, A.V. and PHILLIPS, I. (1989) The preservation of microorganisms in biological specimens stored at −70 °C. *J. Appl. Bacteriol.* **67**, 417–23.

MIGUENS, M.P. (1984) Methods for maintaining stock cultures. *Mykosen* **28**, 134–7.

PERTOT, E., PUC, A. and KREMSER, M. (1977) Lyophilization of nonsporulating strains of the fungus *Claviceps*. *Eur. J. Appl. Microbiol.* **4**, 289–94.

PILET, P.-E. (ed.) (1985) *The physiological properties of plant protoplasts* (Berlin: Springer Verlag).

SHARP, R.J. (1984) The preservation of genetically unstable microorganisms and the cryopreservation of fermentation seed cultures. In: A. Mizrahi, and A.L. van Wezel (eds) *Adv. Biotechnol. Proc.*, Vol. 3, pp. 81–109 (New York: Alan R. Liss).

STAHL, C., NEUMANN, D., SCHMAUDER, H.-P. and GROEGER, D. (1977) Protoplastengewinnung bei Mutterkorn. *Biochem. Physiol. Pflanzen* **171**, 363–8.

STRAUβ, A., DOMASCHKA, G. and SCHMAUDER, H.-P. (1989) Biotechnologische Produktionsverfahren, Lehrbriefe 2 und 3 der Lehrbriefreihe 'Biotechnologie' (Berlin: Zentralstelle für das Hochschulfernstudium).

WEBER, H. (ed.) (1993) *Allgemeine Mykologie* (Jena: Gustav Fischer Verlag).

2.3 PRINCIPLES OF MICROBIAL GENE TECHNOLOGY

M. Schweizer

2.3.1 *Manipulation of Plasmids*

Isolation of Plasmid DNA from *Escherichia coli*

Several methods, all employing the same basic techniques but differing in the amount of time they require and, more importantly, in the degree of purity of the DNA obtained, have been described for plasmid isolation.

Principle

The host system used for the propagation of plasmid DNA is the bacterium *Escherichia coli*. Plasmids are extrachromosomal, double-stranded DNA molecules, usually circular, found in prokaryotes and in some eukaryotes. They vary in size from 2 to >200 kb. Furthermore, plasmids differ in the information they carry, the way in which the DNA is transferred from mother to daughter cell and the copy number per cell. In addition to the genes for DNA replication, plasmids may carry genes whose products confer resistance to various antibiotics or heavy metal ions, encode toxins (e.g. Colicin) or proteins involved in nitrogen fixation, to name but a few examples of plasmid-encoded properties. Worthy of mention is that plasmid-carrying organisms are capable of metabolising aromatic and aliphatic hydrocarbons, and the biotechnological breakdown of phenolic compounds and oils is an important contribution towards a clean, healthy environment. Because of the expression of the above-mentioned genes, i.e. their transcription and translation, the plasmid-containing *E. coli* has a selective advantage over the bacterial cells lacking the plasmid. This opens new prospects for the isolation of such plasmids since they can be selected for on the basis of their phenotype. In the wild this is disadvantageous; under non-selective conditions the plasmid will be lost.

Two methods are used routinely in the laboratory for the isolation of plasmid DNA:

1   Isolation of plasmid DNA on a preparative scale and purification by column chromatography or CsCl density gradient centrifugation.

2   The so-called 'quick prep' on an analytical scale, useful when only the presence or absence of a specific plasmid is to be determined.

Many of the plasmids used routinely in the laboratory are present in the cell in a high copy number and therefore yield a high amount of DNA. Plasmids which are present only in a low copy number, e.g. pBR322, one of the very first plasmids used in molecular biology, must be 'amplified'.

On average there are 15 copies of pBR322 per *E. coli* cell, and upon amplification this number can be increased to 1000–3000 copies per cell. Addition of the antibiotic chloramphenicol to the growing bacterial culture inhibits protein biosynthesis and, as a result, the replication of the chromosomal DNA. Plasmid pBR322 and derivatives thereof, however, replicate their DNA in the absence of *de novo* protein biosynthesis because they possess a Col E1-based origin of replication. Consequently, plasmid DNA accumulates in the cell and can attain levels as high as 40% of the total DNA. In both methods *E. coli* cells are lysed by the addition of detergent, e.g. sodium lauryl sulphate; chromosomal DNA is denatured by treatment with alkali and removed in the form of a salt precipitate. The chromosomal DNA in *E. coli* is a circular molecule of 4700 kb, and even under the most gentle conditions of cell lysis it will undergo fragmentation. Both chromosomal and plasmid DNA are denatured by alkali treatment but following neutralisation with potassium acetate the plasmid, but not the chromosomal DNA, regains its superhelical conformation. The reason is that the two strands of the plasmid DNA cannot be separated because of their supercoiling.

Ribonucleic acid (RNA) is removed by incubation with ribonucleases which recognise specific nucleotide sequences, and protein material is removed by phenol extraction. The phenolised and ribonuclease-treated preparation yields two phases upon centrifugation: the lower organic phase and the upper aqueous phase. The aqueous phase serves as a solvent for the denatured proteins which accumulate in the white interphase. Traces of phenol remaining in the aqueous layer are removed by chloroform or ether extraction. Plasmid DNA is precipitated by either isopropanol or ethanol in the presence of monovalent cations at a final concentration of 0.3 M.

Preparative Isolation of the Low-copy-number Plasmid pBR322 from *E. coli* using the Potassium Acetate Method

Material

LB broth (Luria–Bertani): tryptone 10 g, yeast extract 5 g, NaCl 10 g, NaOH 0.2 g, dH$_2$O ad 1000 ml. *LB + ampicillin* (prior to the addition of ampicillin to a final concentration of 50 mg/ml the medium must be allowed to cool down to approximately 50 °C). *Chloramphenicol*: Chloramphenicol 10% (w/v) in 95% EtOH. A strain of *E. coli* carrying the plasmid pBR322, e.g. RR1{F$^-$ *hsdS20* (r$^-$ m$^-$) *recA$^+$ ara14 proA2 lacY1 galK2 rpsL20 xyl5 mtl1 supE44* λ^- (pBR322)}. 2 litre Erlenmeyer flask; rolling drum; shakers; centrifuges (Beckmann or Sorvall Superspeed RC-5B or equivalent) and the appropriate rotors and tubes (sterile polycarbonate tubes); glass pipettes; automatic pipetting aids; Speed Vac vacuum concentrator. *TE(10/1)* (10 mM Tris-HCl, 1 mM EDTA, pH 8.0). *TE(50/10)* (50 mM Tris-HCl, 10 mM EDTA, pH 8.0). *Lysis buffer* (0.2 M NaOH, 1% SDS (sodium lauryl sulphate)). 3 M *K acetate*, pH 4.8. *Phenol* saturated with TE(10/1). (*Warning!* Phenol is a strong irritant; protective clothing and safety glasses must be worn.) Phenol is heated to 65 °C and then added to an equal volume of TE(10/1) also at 65 °C, stirred thoroughly and cooled to room temperature. After separation of the phases the phenol phase is extracted twice with an equal volume of TE(10/1). This TE-saturated phenol is stored under TE(10/1) in a dark bottle at 4 °C. [Purchasing TE-saturated phenol is safer and less time-consuming.] *RNAse A* (EC 3.1.27.5); (10 mg/ml) in 50 mM sodium acetate, pH 4.8. RNAse A is usually contaminated with DNAse and this contaminating activity can be removed by boiling the RNAse solution for 5 min and cooling it immediately in an ice bath. *RNAse T1* (EC 3.1.27.3) (100 U). 3 M *sodium acetate. Chloroform. Ethanol* (EtOH): 70% (v/v), 95% (v/v). *Isopropanol*. Sterile working conditions must be observed throughout.

Method

1 Inoculate 5 ml LB + ampicillin with a single colony of an *E. coli* strain containing pBR322, and incubate O/N at 37 °C.

2 Inoculate 500 ml LB + ampicillin (in a 2 l Erlenmeyer flask) with 0.5 ml of the O/N culture and incubate, with shaking, at 37 °C until the culture reaches an OD$_{600}$ of 0.6.

3 For amplification of the plasmid, add 1.2 ml of a 10% solution of chloramphenicol in EtOH to the culture, then incubate it with shaking for a further 5–15 h.

4 Harvest the cells by centrifugation in a Sorvall or Beckmann centrifuge (Sorvall JA-14 rotor or Beckmann GSA rotor) for 5 min at 8000 rpm and 4 °C.

5 Pour off the supernatant and resuspend the cell pellet in 5 ml TE(50/10).

6   Lyse the cell suspension by adding 10 ml 0.2 M NaOH / 1% SDS; mix carefully by gently shaking the centrifuge beaker (do *not* use a vortex mixer!).

7   Protein, chromosomal DNA (attached to the cell membrane) and lipids are precipitated following the addition of 5 ml 3 M K acetate, pH 4.8. Mix thoroughly but carefully by swirling while adding the K acetate solution.

8   Centrifuge the milky-white suspension for 20 min at 16 000 rpm and 4 °C. Transfer the clear supernatant, which contains plasmid DNA and RNA, to a sterile centrifuge beaker (for JA-20 rotor) using a 10 ml pipette and noting the volume.

9   Add 0.7 vol of ice-cold isopropanol to the nucleic acid-containing supernatant from step 8. Mix well before centrifuging for 20 min at 16 000 rpm and 4 °C.

10 Dry the pellet in a 'Speed Vac' concentrator and dissolve in 5 ml TE(50/10).

11 Pipette 50 μl RNase A (boiled for 5 min and immediately cooled on ice to destroy any contaminating DNAses) and 1 μl T1-RNAse (100 U) into the solution of nucleic acids and incubate the whole at 37 °C for 1 h.

12 Deproteinise the preparation by extraction with an equal volume of TE-saturated phenol. After vigorous mixing and centrifuging (JA20, 15 min at 14 000 rpm and 20 °C) to separate the phases, transfer the upper aqueous phase to a fresh centrifuge tube and extract with an equal volume of chloroform. Repeat the centrifugation to separate the phases.

13 Precipitate the plasmid DNA contained in the upper phase from step 12 by the addition of 1/10 vol of 3 M sodium acetate and 2.5 vol EtOH (95%, −20 °C). Collect the DNA by centrifugation (JA-20, 16 000 rpm, 4 °C, 20 °C). Rinse the precipitate with 70% EtOH, dry *in vacuo* and dissolve in 1 ml TE(10/1).

With this method the yield is generally about 1–2 mg plasmid DNA / 500 ml bacterial culture.

Isolation of the High-copy-number Plasmid pUC8 using the 'Mini-screen' Method

pUC plasmids are known as 'multi-copy' plasmids, i.e. there are more than 20 copies per bacterial genome in the host cell. The pUC series of cloning vectors contain DNA from the genome of the bacteriophage M13 and the plasmid pBR322 but not that part of the plasmid which is responsible for restricting the copy number to one copy per bacterial genome. Two advantages of these plasmids are the large number of restriction sites, facilitating the insertion of foreign DNA, and the possibility of 'blue/white' selection. The insertion site for foreign DNA in the pUC plasmid is a 'polylinker' inserted into the reading frame of the *lacZ* gene from *E. coli*. Thus it is possible to express fusion proteins which can be detected easily because the cells harbouring such a plasmid produce white not blue colonies on media containing a chromogenic substrate for β-galactosidase. The 'polylinker' is a stretch of DNA corresponding to specific restriction endonuclease sites. The pUC plasmids are 2.7 kb long and can be selected for on the basis of the ampicillin resistance gene they carry.

Material

Plasmid-containing *E. coli* strain, e.g. JM109 (*endA1 recA1 hsdR17* (r⁻m⁻) *thi1 gyr96 relA1 deoR⁺supE44* λ⁻ Δ(*lac-proAB*) F' *traD36 proAB lac I*^{q}*ZΔM15*) transformed with pUC8. *Glucose minimal medium*: K_2HPO_4 10.5 g, KH_2PO_4 4.5 g, glucose 2.0 g, $(NH_4)_2SO_4$ 1.0 g, Na citrate $\times$ 2 H_2O 0.5 g, $MgSO_4 \times 7$ H_2O 0.2 g, thiamine-HCl 5 μg, Bacto agar 20.0 g. To avoid loss of the F'-plasmids, JM strains and RZ1032 (see p. 61) should be stored on minimal medium plates and restreaked at regular intervals. Test tubes containing 3 ml LB broth supplemented with the appropriate antibiotic. Sterile reaction tubes with a volume of 1.5 ml, solutions and equipment as described for the previous experiment.

Method

1 Inoculate 3 ml LB broth + antibiotic with a single colony of an *E. coli* strain containing a plasmid of the pUC series and incubate O/N (8–16 h) at 37 °C.

2 Transfer 1.5 ml of the bacterial culture into a reaction tube and harvest by centrifugation for 30 s in an Eppendorf or similar centrifuge. Pour off the medium or remove by aspiration.

3 Following resuspension of the cell pellet in 0.1 ml TE(50/10), add 0.2 ml 0.2 M NaOH / 1% SDS and 0.15 ml 3 M K acetate, pH 4.8. Mix the contents thoroughly by rocking the tube back and forth several times, and then centrifuge for 5 min (see step 2).

4 Transfer the clear supernatant (by pouring) to a fresh reaction tube. Add 1.0 ml isopropanol (-20 °C), and after mixing the contents by inversion centrifuge the tube for 10 min.

5 Pour off the supernatant and take up the pellet (white) in 0.1 ml TE(10/1): after resuspension, add 0.3 ml 95% EtOH. Mix contents of the tube thoroughly by vortexing and centrifuge for 10 min. Dry the pellet *in vacuo* using a 'Speed Vac' concentrator, and dissolve in 20–50 μl TE(10/1).

The yield of plasmid DNA is 0.03–0.05 mg per 1.5 ml culture. This method lends itself well to the isolation of several plasmids simultaneously; an ideal number is the number of positions available in the centrifuge rotor and, since all tubes contain the same volume there is no need for time-consuming balancing of the rotor.

Purification of Plasmid DNA

The DNA obtained by the above procedure is still contaminated to a certain degree with chromosomal DNA, RNA and protein. These can be removed by column chromatography or CsCl/EtBr density gradient centrifugation.

Column Chromatography

MATERIAL

Glass chromatography column (diam. 2 cm; length 40 cm) and fittings; photometer, Sepharose 4B (Pharmacia); 0.5 M NaCl, 50 mM Tris-HCl, 10 mM EDTA, pH 8.0; TE(10/1); EtOH.

METHOD

1   Prepare a column using Sepharose 4B which has been equilibrated with 0.5 M
    NaCl, 50 mM Tris-HCl, 10 mM EDTA, pH 8.0 and degassed. Once the column
    has been poured, equilibrate it again.

2   Load plasmid DNA dissolved in TE(10/1) onto the column, and as soon as it has
    soaked in, pass 0.5 M NaCl, 50 mM Tris-HCl, 10 mM EDTA, pH 8.0 through
    the column. Adjust the flow rate to approx. 0.35 ml/min and collect fractions of
    7.0 ml. Identify the DNA-containing fractions by recording the OD_{260} using a
    flow-through cuvette. It is also possible to determine OD_{260} 'by hand'.

3   The first OD_{260} peak corresponds to plasmid DNA; pool the fractions with the
    highest values and precipitate the DNA by the addition of one volume of
    isopropanol (place at $-70\,°C$ for 60 min). Collect the plasmid DNA by
    centrifugation at $4\,°C$ and 16 000 rpm for 20 min (JA-20 rotor). Rinse the DNA
    pellet with 70% EtOH and dry *in vacuo*, before dissolving in 0.2–1.0 ml
    TE(10/1). The second peak which is eluted contains RNA and, to a lesser extent,
    chromosomal DNA.

CsCl/EtBr Density Gradient Centrifugation

Separation of supercoiled plasmid DNA molecules from linear and open-circle
plasmid DNA and chromosomal DNA can be achieved by CsCl density gradient
centrifugation in the presence of ethidium bromide (EtBr). EtBr is a phenanthrene
derivative which intercalates between the bases of the DNA and fluoresces orange
when irradiated with UV light. The intercalation of EtBr in the DNA causes a change
in the density of the DNA, allowing the supercoiled plasmid DNA to band at a
position below that of open-circle and linear plasmids and the chromosomal DNA.
Since EtBr intercalates more readily with linear than with circular DNA it causes a
more pronounced reduction in the density of the former. Any remaining RNA
sediments at the bottom of the gradient.

MATERIAL

Ultracentrifuge and Ti50 or VTi65.2 rotors (*pay special attention to the rules
concerning balancing tubes in the ultracentrifuge!*). Quickseal centrifuge tubes of
the appropriate size; sterile disposable 5 ml syringes and needles ('18 gauge');
Ethidium bromide (10.0 mg/ml in TE(10/1)). TE(10/1); CsCl; liquid paraffin; *n*-
butanol saturated with H_2O.

   Extreme care must be exercised when working with EtBr; all solutions must be
decontaminated immediately!

METHOD

1   (i)   *VTi65.2 rotor.* For each DNA sample to be centrifuged, dissolve 3.58 g
          CsCl in 3.38 ml TE(10/1) in a 'Quickseal' polyallomer tube (13×51 mm),
          add 0.2–0.5 ml (maximum 5 mg) plasmid DNA and 0.3 ml EtBr solution.
          After carefully balancing the tubes, seal them as directed by the manufac-
          turer and place them in the rotor. Centrifugation is carried out at $18\,°C$ and
          55 000 rpm for at least 16 h.

    (ii)   *Ti50 rotor*. In either a 'Quickseal' tube or a 50 Ti tube dissolve 7.56 g CsCl in 7.43 ml TE(50/10) containing 0.5 mg DNA and 0.25 ml EtBr solution. Proceed as described in 1(i) for 'Quickseal' tubes. If using capped tubes, overlay EtBr solution with paraffin before sealing. Centrifugation is carried out at 20 °C and 42 000 rpm for 48 h.

2   The bands corresponding to the different species of DNA can be visualised under UV light. Release the vacuum by inserting a needle into the top of the tube ('Quickseal') or removing the screw from the cap of the 50 Ti tube. Remove the lower band, which contains supercoiled plasmid DNA, by inserting an 18 gauge needle attached to a 5 ml disposable syringe just below the band and slowly and steadily withdrawing the band into the syringe. *Caution! EtBr – gloves must be worn.* Avoid removing the upper band which contains 'nicked' plasmid DNA contaminated with chromosomal DNA.

Removal of Ethidium Bromide and Precipitation of Plasmid DNA

METHOD

1   Take up 1 vol of H_2O-saturated *n*-butanol into the syringe containing the plasmid DNA / EtBr mixture. Shake the syringe vigorously and allow it to stand with the needle pointing upwards, until the phases have separated (approx. 1 min). The organic phase, containing EtBr, is expelled from the syringe and discarded (into the organic waste). This process is repeated 4–5 times or until the aqueous phase is no longer coloured red.

2   Transfer the aqueous phase to a centrifuge tube (JA-20) and add 10 vols of TE(10/1) to reduce the final concentration of CsCl. Add 2.5 vols 95% EtOH and precipitate the DNA for 1 h at −70 °C or O/N at −20 °C. Centrifuge as described in step 3 of the purification by column chromatography.

Measurement of Plasmid DNA Concentration

METHOD

The concentration of pure nucleic acid preparations is determined by measuring OD_{260}. The following conversions are used:

- Concentration of plasmid or genomic DNA:

$$OD_{260} \times 50 = \mu g/ml$$

- Concentration of single-stranded DNA or RNA:

$$OD_{260} \times 36 = \mu g/ml$$

- Concentration of oligonucleotides:

$$OD_{260} \times 25 = \mu g/ml$$

To estimate the amount of protein in nucleic acid preparations, OD_{280} is determined and the relative value OD_{260}/OD_{280} calculated; if this lies between 1.8 and 2.0 the preparation is considered to be free of protein.

Further reading

BIRNBOIM, H. and DOLY, Y. (1979) A rapid alkaline extraction procedure for screening recombinant plasmid DNA. *Nucl. Acids Res.* **7**, 1513–23.

BOLIVAR, F., RODRIQUEZ, R.L., GREENE, M.C., BETLACH, M.C., HEYNEKER, H.L., BOYER, H.W., CROSA, J.H. and FALKOW, S. (1977) Construction and characterization of new cloning vehicles, II: A multipurpose cloning system. *Gene* **2**, 95–113.

CLEWELL, D.B. and HELINSKI, D.R. (1972) Nature of ColE1 plasmid replication in *Escherichia coli* in the presence of chloramphenicol. *J. Bacteriol.* **110**, 667–76.

SAMBROOK, J., FRITSCH, E.F. and MANIATIS, T. (1989) *Molecular cloning – a laboratory manual*, 2nd edn, Vol. I, section 1 (Cold Spring Harbor, NY: Cold Spring Harbor Laboratory Press).

YANISCH-PERRON, C., VIEIRA, J. and MESSING, J. (1985) Improved M13 phage cloning vectors and host strains: nucleotide sequence of the M13mp18 and pUC19 vectors. *Gene* **33**, 103–19.

2.3.2 *Restriction Analysis and Agarose Electrophoresis of DNA*

Principle of the Experiment

Restriction endonucleases are an ideal tool for the molecular biologist because they recognise specific sequences in double-stranded DNA and hydrolyse phosphodiester bonds therein. Restriction endonucleases are part of the restriction-modification system of prokaryotes, often referred to as a simple immune system, which recognises and destroys 'foreign' DNA. This system comprises a DNA-modifying and a DNA-cutting mechanism: DNA modification is achieved by DNA methylation which protects the cell's own DNA from attack by restriction endonucleases.

Three classes of restriction endonucleases are known, and it is the *Type II enzymes* which are generally used in cloning of DNA. In the enzymes of type II the endodesoxyribonuclease and the methylase activities are found on separate polypeptides, making them ideal for the isolation of DNA to be cloned into plasmids. The recognition sequence of most restriction enzymes has a palindromic structure, i.e. the nucleotide sequence reads the same in both directions.

More than 800 different restriction enzymes are known. The restriction enzymes are named according to the following rules: an acronym for the organism from which the enzyme was isolated – e.g. *Eco* for *Escherichia coli*, *Hind* for *Haemophilus influenzae* serotype d – followed by a roman numeral. This numbering indicates that several enzymes, each with a different recognition sequence, have been isolated from the same organism: e.g. *Sac*I and *Sac*II. Both enzymes are isolated from *Streptomyces achromogenes* but *Sac*I recognises the sequence GAGCTC whereas *Sac*II recognises CCGCGG.

Specific buffer and temperature conditions are required for the restriction of DNA. In most cases Mg^{2+} ions are essential. Many restriction endonucleases produce DNA fragments with single-stranded overhangs, and these can anneal with the complementary ends of the same DNA or with fragments of foreign DNA having the same or compatible overhangs. Some restriction enzymes, however, cut DNA in such a way that the resulting fragments are blunt-ended, and these can therefore be joined to any other blunt-ended DNA fragments.

The frequency with which the recognition site for a specific restriction endonuclease occurs in any DNA can be determined as follows:

Frequency of restriction sites $= 1/4^N$

where N is the number of base pairs (bp) in the recognition sequence. Therefore, a restriction enzyme with a tetrameric recognition site will cut once every 256 bp. On the other hand if a restriction enzyme recognises a hexameric sequence it will cut the DNA only once every 4096 bp.

The second technique described in this section is electrophoresis in agarose gels. Agarose consisting of polymerised galactose residues dissolves on heating to give a clear solution which can be poured into a gel casting form and upon cooling solidifies to give a gel matrix. The density of the gel matrix is determined by the initial agarose concentration. For the separation of DNA fragments obtained by restriction enzyme treatment, the concentration of agarose should lie between 0.4% and 1.5%. In an electric field, DNA, which is a polyanion, migrates through the gel matrix towards the anode. The length of time required to separate DNA fragments by electrophoresis is dependent on a number of parameters, e.g. agarose concentration, volume and composition of the buffer. The distance travelled by a linear DNA fragment is inversely proportional to $\log_{10}$ of the size of the fragment in kb or dalton (g/mol). The two dyes, xylene cyanol and bromophenol blue can be used as 'markers' for the migration of DNA fragments of 0.7–9.0 kb. For instance, in a 0.8% agarose gel the 'ideal' positions of the two marker dyes at the end of the electrophoresis are at 1/3 and 2/3 of the total length of the gel.

Description of the Experiment

As a general rule, plasmid DNA is incubated with 1–2 units of a restriction endonuclease per µg DNA using the buffer and temperature conditions recommended for the enzyme by the manufacturer. One unit (U) of restriction endonuclease is that amount of enzyme which totally digests 1 µg substrate DNA (usually bacteriophage λ DNA) in 1 h at 37 °C under optimal buffer conditions. Many restriction enzymes can be inactivated by heating to 65 °C for 10 min.

For the experiment described below one can use plasmids which are readily available in the laboratory (e.g. pBR322, pUC8, pBluescript KS(+), etc.). The DNA should be digested not just with a single enzyme but also with two or three enzymes in combination. For double and triple digests TAM buffer as described by O'Farrell *et al.* (1980) is especially useful. However, if only partial digests are obtained in TAM buffer it will be necessary to adjust the buffer conditions for each individual enzyme. In this case the digest should be carried out first with the enzyme requiring no or low salt in the reaction buffer. After 1–2 h incubation the salt concentration is adjusted to the molarity required for the second enzyme, the second enzyme is added and the incubation continued for a further 2 h. Information about the activities and 'star' activities of restriction enzymes can be found in the pamphlet 'Promega Notes' No. 36. Pharmacia provide a One-phor-all *plus* buffer in which most restriction enzymes show optimal activity. The catalogues of the companies supplying molecular biologicals (e.g. Boehringer Mannheim, New England Biolabs, Life Technologies, Stratagene, Pharmacia) are excellent sources of information on restriction enzymes.

Digestion of Plasmid DNA with Various Restriction Endonucleases and Separation of Products in Agarose

Plasmids

pBR322. This is an *E. coli* cloning vector which carries genes conferring resistance to ampicillin and tetracycline and an origin of replication. The vector has a length of 4.36 kb, and within the ampicillin and tetracycline genes there are a number of recognition sites for restriction endonucleases. For example, *Pst*I cuts within the ampicillin resistance gene whereas *Bam*HI, *Hin*dIII, *Sal*I and *Cla*I all cut within the tetracycline resistance gene. These genes are therefore suitable as selection markers in cloning experiments.

pUC8. See pBluescriptKS(+).

pBluescriptKS(+). A 2.9 kb derivative of the pUC plasmid series, it contains a multiple cloning site (polylinker) with recognition sites for 21 restriction endonucleases. The recognition sites are so arranged that those for enzymes creating 5'-overhangs are flanked on both sides by recognition sites for endonucleases creating 3'-overhangs. Furthermore, the presence of T3 and T7 promoters on either side of the polylinker allows *in vitro* transcription with either T3 or T7 RNA polymerases. Yet another advantage of this plasmid is that it carries an intergenic sequence of M13 permitting the isolation of single-stranded DNA with the aid of a suitable helper phage.

KS(+) describes the orientation of the polylinker region (*Kpn*I → *Sac*I) with respect to the direction of transcription of the *lacZ* gene. The vector pBluescriptKS(+) and the pUC series of vectors contain part of the *lacI* gene and the first 492 nucleotides of the *lacZ* gene (*lacIOPZ'*) as well as the gene-encoding β-lactamase (amp^r) and an origin of replication from *E. coli*. The polylinker is inserted into the *lacZ* gene so that the reading frame is not affected.

The enzyme which is encoded by the *lacZ* gene of *E. coli* plays an important role in molecular biology (blue/white selection). The inactivation of the *lacZ* gene by insertion of foreign DNA can be detected by a change in colony colour. The α-peptide of β-galactosidase (= the first 146 amino acids of β-galactosidase) is enzymatically inactive, but it can aggregate with another likewise inactive fragment of the enzyme lacking amino acids 11−41 to yield an active complex. This phenomenon is known as α-complementation and depends on the non-covalent association of peptide fragments, which allows two inactive peptides to aggregate in such a way that an active β-galactosidase is created. This property of β-galactosidase is used as the selection system for pBS and pUC vectors. For selection purposes the plasmid carries the coding information for the α-peptide, and the bacterial strain used as the host contains a so-called F' plasmid (fertility factor) on which the *lacZ* gene containing the M15 deletion (spanning the codons for amino acids 11−41) is located. In the presence of the chromogenic substrate for β-galactosidase X-Gal (5-cromo-4-chloro-3-indolyl-β-galactoside), bacterial colonies expressing β-galactosidase are blue in colour. However, if the DNA region encoding the α-peptide is interrupted by the insertion of foreign DNA, α-complementation cannot occur, and as a consequence no active β-galactosidase is present in the cell. These cells containing recombinant plasmids give rise to white colonies on X-Gal plates, hence the term 'blue/white' selection.

Electrophoresis chambers (75×75 mm or 225×115 mm) and accessories; Polaroid MP 4 Land camera; power supply; UV transilluminator; sterile reaction tubes. Agarose (e.g. Ultrapure, Life Technologies). $5 \times$ TBE (Tris-base O.45 M, boric acid O.45 M, EDTA 0.125 M, pH 8.3; the pH of the buffer is automatically 8.3 when the substances have dissolved but checking is advisable); $1 \times$ TBE ($5 \times$ TBE diluted $1:5$ with dH$_2$O). *Loading buffer* (glycerol 59%, bromophenol blue 0.03%, xylene cyanol 0.03%). *DNAs*: bacteriophage λ DNA, vectors (pBR322, pUC series, pBluescript KS($+$) from Stratagene, Life Technologies, Boehringer Mannheim, BioLabs, etc. *Restriction endonucleases*: e.g. *Eco*RI, *Bam*HI, *Pst*I, *Hind*III, *Pvu*I, *Sma*I etc. (see Appendix I, 'Restriction endonuclease targeting sites', pp. 371–3 in *Principles of gene manipulation*, Old and Primrose (1989)). *10 × TAM (O'Farrell buffer)* (K acetate 0.66 M, Tris-HCl 0.33 M, Mg acetate 0.1 M, dithiothreitol 5 mM, bovine serum albumin 100 mg/ml, pH 7.9). *Ethidium bromide* (see above).

1   Restriction enzyme digest: pipette the following into a sterile reaction tube:

μl DNA ($1–2$ μg)

2 μl 10 × TAM (the final concentration of the buffer is 1 ×)
μl enzyme ($2–5$ U)

$\underline{\mu l\ H_2O}$

20 μl

Do not forget a control tube! One possible control is a tube from which the restriction enzyme is omitted. When this sample has been run on the gel several bands will be visible: the lowest one corresponds to the supercoiled form of the plasmid which has migrated furthest into the gel; the band immediately above this is the 'open' circle form. Any bands higher up in the gel are due to concatameric forms of the plasmid. In some plasmid preparations it is possible that 'shearing' of the DNA has occurred, and in this case bands corresponding to linear plasmids appear between the supercoiled (closed circle) and 'open' circle plasmid species.

(i)   Digest of plasmid DNA with several restriction enzymes (e.g. *Hind*III + *Eco*RI; *Hind*III + *Bam*HI; *Hind*III + *Pst*I, etc.).

(ii) λ DNA as standard cut with *Hind*III or *Hind*III/*Eco*RI. An aliquot of one or both of these standards will be loaded onto the gel to allow the sizes of the restriction fragments obtained with the plasmid DNAs to be determined. The lengths of the fragments (in kb) of λ DNA obtained are:

(a)   *Hind*III-restricted λ DNA: 23.1, 9.4, 6.5, 4.3, 2.3, 2.0, 0.56, 0.125.

(b)   *Hind*III/*Eco*RI-restricted λ DNA: 21.2, 5.1, 4.9, 4.3, 2.1, 1.9, 1.6, 1.3, 0.98, 0.83, 0.56, 0.125.

2   Incubate the samples at 37 °C for $1–2$ h.

3   Stop the reaction by placing the samples at 65 °C for 10 min.

4   Add 1/10 vol loading buffer to each sample and place again at 65 °C for 10 min (this second incubation at the higher temperature is particularly important for the

λ DNA digest to disrupt any concatemers which may have formed), and keep samples on ice until the gel is to be loaded.

5   Subject the samples to electrophoresis at 40–120 V either in horizontal 'Minigels' (75 × 75 mm) with 14 wells or in larger gels measuring 225 × 155 mm with 14 or 32 wells. The buffer used for the electrophoresis is 1 × TBE, and the agarose concentration is 0.8% in 1 × TBE.

6   Dissolve the agarose by heating on a Bunsen burner or in a microwave oven. Allow the agarose to cool to approx 55 °C before pouring into the gel form and inserting the comb to form the wells. The gel should be about 0.5 cm in depth.

7   After the gel has set (the process can be speeded up by placing the gel in the refrigerator or cold room) the comb is removed carefully and the gel is placed gently in the electrophoresis chamber. Pour in sufficient 1 × TBE to cover the gel to a depth of 2–3 mm.

8   When the electrophoresis has been completed it is placed in a 0.1% solution of EtBr for 5 min to stain the DNA in the gel. (*Safety! Wear disposable gloves.*) The DNA is then visualised by exposing it to UV light. This is done by placing the gel on a transilluminator. The result of the electrophoresis is documented by photographing the gel in UV light using a MP 4 Land camera and Polaroid 667 film. (*Safety! Wear a protective face shield when using the transilluminator.*)

Restriction Mapping

The determination of the positions of restriction sites in the plasmid DNAs used in the restriction digests described above is termed restriction mapping and can be carried out as follows. The plasmid DNA is digested with various restriction enzymes and the fragments obtained separated by electrophoresis as described. With the help of appropriate double digests the positions of the restriction sites can be determined. The lengths of the DNA fragments obtained can be determined from the standard curve drawn by plotting the $\log_{10}$ of the distance migrated for each of the λ DNA restriction fragments obtained with *Hind*III and/or *Hind*III/*Eco*RI against the length of the fragment.

Decontamination of EtBr-containing Solutions

Ethidium bromide is a powerful mutagen, and therefore disposable gloves must always be worn when handling any solutions containing it. On no account should EtBr-containing solutions be disposed of by pouring down the sink.

Decontamination of Used Electrophoresis Buffer Containing ca. 0.5 μg/ml EtBr

1   Add 1.0 g active charcoal per 1.0 l EtBr-containing buffer.

2   Mix thoroughly for 1 h.

3   Filter the suspension through Whatman paper No. 1 and discard the filtrate. The active charcoal can be reused several times.

4   Label and dispose of the used active charcoal and filters as recommended for halogen-containing waste.

References and further reading

BENSAUDE, O. (1988) Ethidium bromide and safety – readers suggest alternative solutions. *Trends Genet.* **4**, 89.

DAVIS, C.C. and DILNER, M.D. (1986) *Basic methods in molecular biology* (Amsterdam: Elsevier).

KAPPELMANN, J. (1992) Troubleshooting restriction enzyme digestions. In: *Promega Notes* No. 36, p. 16.

O'FARRELL, P.H., KUTTER, E. and NAKANSKI, M. (1980) A restriction map of the bacteriophage T4 genome. *Mol. Gen. Genet.* **179**, 421–35.

OLD, R.W. and PRIMROSE, S.B. (1989) *Principles of gene manipulation – an introduction to genetic engineering*, 4th edn (London: Blackwell Scientific Publications).

Promega Protocols and Applications Guide, 3rd edn (1996) (Heidelberg: SERVA).

WILSON, G.G. and MURRAY, N.E. (1991) Restriction and modification systems. *Ann. Rev. Genet.* **25**, 585–627.

WINNACKER, L.E. (1990) *Genes and clones* (Weinheim: Verlag Chemie).

2.3.3 Ligation of DNA Fragments

Principle of the Experiment

The joining of DNA fragments is the 'secret' of all gene technology experiments and is the basic step in the cloning of genes. The process is carried out *in vivo* and *in vitro* by DNA ligases, enzymes which catalyse the formation of a phosphodiester bond between a free 3'-hydroxyl group and a 5'-phosphate group. Ligases play an important role in recombination, replication and repair of DNA and they have been isolated from a number of prokaryotes and eukaryotes, as well as viruses. In the case of *E. coli* DNA ligase NAD^+ is required as a co-substrate for the formation of an enzyme–adenylate complex prior to the formation of the phosphodiester bond. In contrast, DNA ligase of the bacteriophage T4 requires ATP for the formation of such a complex. Upon formation of the phosphodiester bond, NMN^+ or AMP is released.

In the simplest case the construction of a recombinant plasmid is a bimolecular reaction in which the free ends of a linearised vector DNA are joined with the free ends of the DNA fragment to be cloned. Therefore one can adjust the conditions of a ligation reaction such that the formation of circular vector molecules, linear oligomers, or cloning of the desired DNA fragment is favoured. The ligation reaction is a complicated process influenced by several parameters, e.g. temperature, ionic strength of the buffer, the type of free DNA ends (5'- or 3'-overhangs or blunt ends), the relative concentration of DNA ends and the molecular weight of the DNA. The temperature at which the cohesive ends produced by the restriction enzyme anneal is approximately 15 °C, and it also depends on their length and sequence. However, DNA ligase has a temperature optimum of 37 °C, and therefore the temperature at which the ligation is carried out has to be a compromise between these two values. As a general rule DNA ligations are performed at 12.5 °C.

Description of the Experiment

The basic theory of DNA ligation will not be described here, but the reader is referred to the article of Dugaiczyk *et al.* (1975). In general one should ensure that the self-ligation of the vector is kept to a minimum. This is usually achieved by having a higher molar concentration of ends of the fragment to be cloned than of the ends of the vector (ratio of molarity of the DNA to be inserted : the molarity of the vector = 1 : 6).

In the experiment described here, *Hind*III restriction fragments of λ DNA are inserted into pUC8 cleaved with *Hind*III, and *Bam*HI-restricted pBR322 DNA is ligated with *Bam*HI-restricted λ DNA. Possible other combinations are, e.g. total genomic DNA from *E. coli* cleaved with *Hind*III inserted into *Hind*III-restricted pUC8 and pBR322. A ligation with blunt-ended restriction fragments will also be described.

Ligation of Plasmid DNA with λ DNA Restriction Fragments

Material

10 × Ligase buffer (Tris-HCl 0.1 M, MgCl$_2$ 0.1 M, dithiothreitol (DTT) 0.01 M, bovine serum albumin (BSA) 1.0 μg/ml, pH 7.5); 10 mM ATP in 10 mM Tris-HCl, pH 7.5; *T4 DNA ligase* (EC 6.5.1.1); 4% PEG 6000 (polyethylene glycol); EDTA 0.2 M, pH 8.0; Tris-HCl 50 mM, spermidine 1 mM, EDTA 0.1 mM, pH 7.6; alkaline phosphatase (calf thymus; EC 3.1.3.1); phenol saturated with TE(10/1); Tris-HCl 50 mM, pH 8.0; column with immobilised phosphatase (Mobitec); EtOH.

The efficiency of the ligation will be checked by electrophoresis. In the ligated sample the linearised plasmid and λ DNA fragments should no longer be visible. As a result of the ligation, bands with a higher molecular weight than that of the restriction fragments should appear. When the ligation has been completed the aliquots will be transformed into *E. coli* (see below) and the transformants analysed as follows:

1 'blue/white' selection

2 inactivation of tetracycline resistance

3 isolation of plasmid DNA from the transformants.

Method

1 The appropriate amounts of DNA restriction fragments to be ligated are mixed with 1/10 vol 10 × ligase buffer in a volume of 10–30 μl. If the restriction digest was carried out in 10 × TAM buffer the ligation reaction can be set up by merely adding ATP and ligase. Another possibility is to precipitate the DNA (addition of 1/10 vol 3 M Na acetate pH 4.8 and 2.5 vol EtOH (95%, −20 °C) and washing with 70% EtOH). Dissolve the DNA in an appropriate volume of sterile ddH$_2$O. Ligation sample: addition of 1/10 vol 10 × ligase buffer, H$_2$O, ATP and T4 DNA ligase.

   (i) For samples for transformation for 'blue/white' selection, controls are:

     (a) pUC8 DNA (unrestricted)

     (b) pUC8 DNA cleaved with *Hind*III

     (c) λ DNA cleaved with *Hind*III

   and ligation samples are:

     (a) pUC8 DNA, *Hind*III restricted and religated

     (b) pUC8 DNA, *Hind*III restricted, religated and cut again with *Hind*III

(c) *Hind*III restricted pUC8 DNA (0.5 μg) ligated with different amounts (0.5 μg, 1 μg, 2 μg, 4 μg, 5 μg) of λ DNA cleaved with *Hind*III

(d) λ DNA cut with *Hind*III and religated (for electrophoresis).

(ii) For samples for transformation for amp^r/tet^s selection, controls are:

(a) pBR322 DNA (unrestricted)

(b) pBR322 DNA cleaved with *Bam*HI

(c) λ DNA cleaved with *Bam*HI (5 fragments)

and ligation samples are:

(a) pBR322 DNA, *Bam*HI restricted and religated

(b) pBR322 DNA, *Bam*HI restricted, religated and cut again with *Bam*HI

(c) *Bam*HI-restricted pBR322 DNA ligated with different amounts of *Bam*HI-restricted λ DNA (see above)

(d) λ DNA cut with *Bam*HI and religated (for electrophoresis).

2 Before the addition of DNA ligase and ATP the samples are incubated at 65 °C for 10 min, then at 37 °C for 10 min and 10 min at room temperature before placing on ice. This procedure enhances the annealing of the cohesive ends.

3 (i) Following the addition of 1/10 vol 10 mM ATP and 1–2 U T4 DNA ligase, the samples are incubated at 12–14 °C for 10–14 h. The reaction is stopped by incubation at 70 °C for 10 min. The samples can either be used immediately for transformation or stored at −20 °C.

(ii) *Ligation of blunt-ended DNA restriction fragments* is carried out in a similar fashion but in the presence of 0.4% PEG 6000 and with 2–4 h incubation at room temperature rather than 14 °C. This procedure can also be adopted for ligation of restriction fragments with cohesive ('sticky') ends. Ligations carried out in this way cannot be stored at −20 °C because DNA is precipitated at low temperatures in the presence of PEG.

Example. Ligate *Sma*I-cleaved pUC8 DNA with *Eco*RV-restricted λ DNA. Do not forget the controls!

Electrophoresis of the Ligation Samples

Aliquots containing 0.5–1.0 μg DNA are withdrawn from each sample and separated by electrophoresis on an 0.8% agarose gel.

Alkaline Phosphatase Treatment of DNA Restriction Fragments

To prevent recircularisation of the linearised plasmid the 5'-phosphate groups may be removed by treatment with alkaline phosphatase. Two methods are described here.

Example. Cleave pUC8 DNA with *Hind*III. The phosphatase-treated DNA can be visualised after electrophoresis and the efficiency of the phosphatase treatment checked by self-ligation of the sample.

Method 1

1   5 μg DNA linearised with the appropriate restriction enzyme is taken up in 50–300 μl of the following 'cocktail' which has a pH of 7.6:

- Tris-HCl   50 mM
- spermidine   1 mM
- EDTA   0.1 mM

and incubated with 1 U calf thymus alkaline phosphatase at 37 °C for 30 min.

2   The reaction is stopped by the addition of 1/10 vol 0.2 M EDTA, pH 8.0 and incubating at 70 °C for 10 min.

3   The sample is extracted with 1 vol TE(10/1)-saturated phenol and the DNA is precipitated with 1/10 vol 3 M Na acetate and 2.5 vol EtOH (95%, at −70 °C) for 1 h. The DNA precipitate is rinsed with 70% EtOH, dried *in vacuo* and dissolved in 20 μl TE(10/1).

Method 2

The 5'-phosphate groups can be more quickly and efficiently removed using phosphatase immobilised on a column (Mobitec 90).

1   The column is rinsed, using a syringe, with 10 ml of sterile 50 mM Tris-HCl, pH 8.0 and then spun for 5 s in an Eppendorf centrifuge.

2   Up to 30 μg restricted DNA in a volume of 40 μl is loaded onto the column, and the whole is incubated at 37 °C for 25 min.

3   The column is placed in an Eppendorf tube and the dephosphorylated DNA is eluted into the tube by centrifuging for 5 s at 3500 rpm.

4   The eluted DNA can be used immediately for ligation purposes or stored at −20 °C.

As a control the dephosphorylated DNA should be ligated, subjected to electrophoresis and transformed into *E. coli* to estimate the efficiency of dephosphorylation.

References

DUGAICZYK, A., BOYER, H.W. and GOODMAN, H.M. (1975) Ligation of EcoRI endonuclease-generated DNA fragments into linear and circular structures. *J. Mol. Biol.* **96**, 171–84.

SAMBROOK, J., FRITSCH, E.F. and MANIATIS, T. (1989) *Molecular cloning – a laboratory manual*, 2nd edn, Vol. I, section 1 (Cold Spring Harbor, NY: Cold Spring Harbor Laboratory Press).

2.3.4   *Transformation of* Escherichia coli *by the CaCl$_2$ Method*

Principle of the Experiment

Transformation may be defined as the introduction of DNA into a suitably prepared prokaryotic cell. The success of many molecular biological experiments is totally dependent on the ability of *E. coli* to take up exogenous DNA. In most instances

$CaCl_2$ treatment, which alters the structure of the cell wall, is used to facilitate the uptake of DNA. After the preparation of competent cells the uptake of DNA is improved by a 'heat shock' which destabilises the lipids in the cell wall. In principle three steps are necessary for the introduction of DNA into *E. coli*:

1   preparation of competent cells

2   transformation of competent cells

3   selection of transformants.

The efficiency of transformation can be affected by different parameters, e.g. the size of the plasmid, the conformation of the DNA, etc. The average frequency of transformation is $10^6 - 10^7$ transformants per μg plasmid DNA.

Description of the Experiment

Both plasmid DNA and M13 DNA will be transformed into *E. coli*.

The bacteriophage M13 is used mainly in gene technology for the production of single-stranded DNA for enzymatic sequencing of DNA using the method of Sanger. Further applications in which M13 phage is used are cloning and *in vitro* mutagenesis. The life cycle of the bacteriophage is exploited for these applications. M13 DNA and M13-derived vector DNA exist in the phage as a single-stranded circle which, after entering *E. coli*, is converted into a doubled-stranded molecule. The newly synthesised complementary strand, also designated 'minus' strand, serves as the matrix for the synthesis of further copies of the double-stranded replicative form of M13 DNA. Double-stranded M13 DNA can be isolated from *E. coli* in the same way as plasmid DNA. Single-stranded DNA, however, occurs later in the life cycle, *viz.* when the infectious phage particles are released into the medium. The bacterial cells are not lysed upon release of the phage particles, but their growth is slowed down. The 'turbid' plaques which arise correspond to areas of reduced bacterial growth. The vectors M13mp18 and M13mp19 are derivatives of the DNA of the filamentous phage M13. M13mp18 and M13mp19 contain the same multiple cloning site which is, however, inserted in the opposite orientation with respect to the vector DNA. When DNA fragments are incorporated into the cloning site the reading frame of the β-galactosidase gene is interrupted, giving rise to a product which, in the presence of X-Gal and IPTG (iso-propyl-β-thiogalactoside is a gratuitious inducer of the *lac* operon which removes the repressor and thereby allows the production of β-galactosidase), does not produce a blue colour and therefore the colonies carrying plasmids with cloned fragments are white not blue; hence the term blue/white selection. Both double-stranded and single-stranded DNA required for sequencing can be isolated from the filamentous phage M13 (see *in vitro* mutagenesis, Section 2.3.9).

Transformation of Plasmids and Recombinant DNA into *E. coli* and Characterisation of the Transformants Obtained

Material

LB broth (see Material in Subsection 2.3.1); *2 × TY broth* (tryptone 16.0 g, yeast extract 10.0 g, NaCl 5.0 g, dH_2O ad 1000 ml); *LB agar* (LB broth + 20.0 g Bacto-agar/1); *LB + amp agar* (LB agar + 50 μg/ml ampicillin); *LB + tet agar* (LB

48

agar + 50 μg/ml tetracycline); *H agar* (tryptone 10.0 g, NaCl 8.0 g, Bacto-agar 15.0 g, dH$_2$O ad 1000 ml; *H top agar* (as *H agar* but with 8.0 g Bacto-agar/l; this 'soft agar' can be melted by heating at 90 °C for 5 min or by heating in a microwave oven); *0.1 M IPTG* (isopropyl-β-thiogalactoside) in ddH$_2$O; *2% X-Gal* (5-bromo-4-chloro-3-indolyl-β-galactoside) made up in dimethyl formamide; 50 mM CaCl$_2$; 100 mM CaCl$_2$ in 15% glycerol; 100 mM MgCl$_2$; dimethyl sulphoxide. Sterile velvets; replica block; sterile reaction tubes; sterile toothpicks or cocktail sticks. RRI[(no pBR322 plasmid!) F$^-$ *hsdS20* (r$^-$m$^-$) *recA*$^+$ *ara14 proA2 lacY1 galK2 rpL20 xy15 mtl1 supE44* λ^-]; JM109[(no pUC 8 plasmid) *endA1 recA1 hsdR17* (r$^-$m$^+$) *thi1 gyr96 relA1 deoR*$^+$ *supE44* λ^- Δ*(lac-proAB)* (F' *traD36 proAB lacI*^{q} *Z*Δ*M15*)].

The ligation samples, prepared as described in Subsection 2.3.3., and the plasmids used to obtain them should be transformed into *E. coli*. The *phenotypic characterisation* of the transformants obtained with pUC plasmids and derivatives thereof is dependent on blue/white selection, whereas the pBR322 transformants are selected on the basis of their antibiotic resistance genes. DNA will be isolated from the transformants and examined by restriction digests and gel electrophoresis. In addition the DNA can be analysed by 'Southern' transfer and hybridisation with a non-radioactively labelled probe and by sequencing. Yet another possibility is 'sub-cloning', i.e. a cloned fragment will be cut by restriction endonucleases into smaller fragments which can then be recloned. There is no need to prepare competent cells for each transformation experiment; a method for the preparation and storage of competent cells is described below. Transformation starting from an O/N culture of *E. coli* is carried out as described below.

Method

1   50 ml LB broth (or 2 × TY broth for transformation with M13-based vectors) is inoculated with 0.1–0.5 ml of an O/N culture of the *E. coli* strain JM109 and incubated, with shaking, at 37 °C until the mid-logarithmic phase of growth is reached.

2   (i)   The culture is divided between two JA-20 centrifuge tubes and the cells are collected by centrifugation at 4 °C and 8000 rpm for 5 min, resuspended in 10 ml ice-cold 50 mM CaCl$_2$ and held on ice for 20 min.

   (ii)   If an M13 transfection is being carried out, add 25 ml fresh 2 × TY broth to the Erlenmeyer flask containing the cells and incubate further at 37 °C so as to have a supply of fresh cells for plating out.

3   The cells are centrifuged a second time, the supernatant discarded and the cell pellet resuspended in 2 ml ice-cold 50 mM CaCl$_2$. 300 μl aliquots are pipetted into sterile reaction tubes, and 1–20 μl of the transforming DNA is added.

4   After 40 min incubation on ice the samples are placed for 3 min in a water bath at 42 °C and then on ice again.

5   (i)   0.5 ml LB broth is added to each of the samples transformed with plasmid DNA, and the tubes are incubated for 1 h at 37 °C (without shaking) to allow the cells to regenerate.

      (a)   *pUC samples*. Collect by centrifuging for 30 s, remove all but approx. 100 μl of the supernatant, add 30 μl X-Gal and 30 μl IPTG to each sample. Mix to resuspend the cells and plate onto selective medium (LB + amp plates). Incubate the plates O/N at 37 °C.

Generally, blue colonies contain only vector DNA, whereas 'white' or transparent colonies contain an 'insert' in the polylinker of their plasmid: in this case either a fragment of λ DNA or a second pUC vector molecule.

(b) *pBR322 samples.* No addition of X-Gal or IPTG is necessary. Harvest the cells as described for the pUC samples and plate onto LB + amp plates. Incubate the plates O/N at 37 °C. If the plates have too many colonies to allow a good 'replica' to be made, the colonies should be picked, using sterile toothpicks, onto fresh LB + amp plates and incubated O/N. On the following day the LB + amp plates are replicated to LB + tet.

Colonies growing on LB + amp but not on LB + tet are inoculated into sterile test tubes containing 2 ml LB + amp after incubation at 37 °C, their DNA is extracted and analysed.

(ii) For M13 transfection add 200 μl of the cells from step 2(ii), 30 μl X-Gal and 30 μl IPTG to each aliquot. Mix thoroughly and add to 3 ml H-top agar held at 45 °C, disperse the cells in the agar by 'rolling' the tube between the palms of your hands for 20–30 s – *avoid producing bubbles!* – and pour gently onto H-plates warmed to 37 °C. When the agar has solidified the plates are inverted and incubated O/N at 37 °C. The transformation rate for each aliquot can be calculated as follows:

$$\text{Transformation rate} / \mu\text{g DNA} = (\text{No. of colonies/plate}) \times \text{total vol.}$$
$$\times \text{ dilution factor} \div \mu\text{g DNA}$$

If a high transformation rate is not required the procedure can be speeded up by incubating the cells with DNA for 10 min and, because cells treated with $CaCl_2$ generally exhibit a good survival rate, the 60 min 'regeneration' in LB broth can be omitted.

Preparation and Storage of Competent *E. coli* Cells

1 Inoculate 100 ml LB broth with 1 ml O/N culture of either JM109 or RR1 and incubate, with shaking, at 37 °C until mid-log phase has been reached.

2 Place the flask on ice before harvesting the cells by centrifuging at 4 °C in JA-14 tubes at 5000 rpm for 5 min; after decanting the supernatant resuspend the cell pellet in ice-cold 100 mM $MgCl_2$.

3 Centrifuge the cell suspension as in step 2. Resuspend the cell pellet gently in 30 ml ice-cold 100 mM $CaCl_2$ and place on ice for 20 min.

4 Harvest the cells as described in step 2 but at 3000 rpm rather than 5000 rpm. Resuspend the cell pellet in 5 ml ice-cold 100 mM $CaCl_2$ / 15% glycerol and dispense in 200 μl aliquots into pre-cooled reaction tubes. The competent cells can be stored at -70 °C.

5 For transformation add 0.1 μg plasmid DNA or up to 10 μl of a ligation sample to 200 μl competent cells which have been thawed gently and placed on ice.

6 After mixing gently, leave the samples standing on ice for 20 min.

7 Place the samples at 42 °C for 2 min and then on ice for a further 5 min incubation before adding 1 ml LB broth and incubating the samples at 37 °C without shaking for 90 min.

8 Centrifuge the samples briefly to collect the cells. Pour off the supernatant and remove any remaining liquid with the aid of an automatic pipette. Take up the cell pellet in 100–200 μl LB, and plate the resulting suspension onto LB + amp selective plates for O/N incubation at 37 °C.

Bibliography

AUSUBEL, F.M., BRENT, R., KINGSTON, R.E., MOORE, D.D., SEIDMAN, J.D., SMITH, J.A. and STRUHL, K. (1987) *Current protocols in molecular biology* (New York: Wiley).

CHUNG, C.T., NIEMELA, S.L. and MILLER, R.H. (1989) One-step preparation of competent *Escherichia coli*: Transformation and storage of bacterial cells in the same solution. *Proc. Natl. Acad. Sci. USA* **86**, 2172–5.

HANAHAN, D. (1983) Studies on transformation of *E. coli* with plasmids. *J. Mol. Biol.* **166**, 557–80.

MESSING, J. (1983) New M13 vectors for cloning. *Methods Enzymol.* **101**, 20–78.

SANGER, F., NICKLEN, S. and COULSON, A.R. (1977) DNA sequencing with chain terminating inhibitors. *Proc. Natl. Acad. Sci USA* **84**, 4767–71.

2.3.5 *Isolation of Chromosomal DNA from* Escherichia coli

In this Subsection the isolation of chromosomal DNA from *E. coli* will be described. Manipulation of the chromosomal DNA should be carried out only in wide-bore glass pipettes. The DNA may be analysed by restriction enzyme digests and gel electrophoresis, and the fragments obtained may be cloned into appropriately restricted pUC8 or pBR322 plasmids.

Material

TES (Tris-HCl 30 mM, EDTA 5 mM, NaCl 50 mM, pH 8.0); *TES/sucrose* (20% sucrose in TES); *EDTA* 0.25 M, pH 8.0; *lysozyme solution* (4 mg/ml lysozyme dissolved in 50 mM Tris-HCl, 10 mM EDTA, pH 7.5 to be made up immediately prior to use); *SDS* (10% w/v); *proteinase K* (EC 3.4.21.14) (10 mg/ml in TE(10/1); *phenol* (saturated with TE(10/1)); *PCI* (phenol/chloroform/isoamyl alcohol [saturated with TE(10/1)] in the ratio 25:24:1); *PC* (phenol/chloroform [TE(10/1) saturated] in the ratio 1:1); *ether*; *RNAse A* (EC 3.1.27.5) (10 mg/ml in 50 mM Na acetate, pH 4.8; boil for 10 min and cool immediately on ice before use); *Na acetate* 3 M; EtOH; sterile glassware and centrifuge tubes; *E. coli* strains (see Subsection 2.3.4).

Method

1 100 ml LB broth, contained in a 1 l Erlenmeyer flask, is inoculated with 0.4 ml of an O/N culture and incubated O/N, with shaking, at 37 °C.

2 The cells are harvested by centrifugation and the cell pellet is resuspended, with the aid of a pipette or by using a magnetic stirrer, in 100 ml TES/sucrose. The suspension is centrifuged again to remove any remaining medium.

3   The cell pellet is resuspended in 10 ml TES/sucrose. Following the addition of 1 ml 0.25 M EDTA, pH 8.0, and 0.5 ml lysozyme solution, the suspension is incubated on ice for 10–15 min to allow spheroplast formation. Spheroplasts are cells whose walls are no longer intact after the lysozyme treatment; the appearance of the spheroplasts can be followed in the microscope.

4   Lysis of the spheroplasts is brought about by the addition of 1/10 vol 10% SDS, gentle but thorough mixing and incubation at 30 °C until a clear viscous solution is obtained.

5   Following the addition of 0.4 ml proteinase K the sample is incubated for a further 30 min at 30 °C.

6   The suspension is deproteinised by extraction with an equal volume of PCI. Shake thoroughly!

7   The phases are separated by centrifugation at 20 °C and 10 000 rpm for 5 min (JA-20 rotor). The lower organic phase and the white interphase are removed by putting a glass pipette through the aqueous and organic phases to the bottom of the centrifuge tube, withdrawing the two lower phases into the pipette with the help of a suction device. An equal volume of PC is added to the aqueous phase, and the extraction procedure described above is carried out again. The extraction with PC is repeated until the aqueous phase becomes clear. If the solution from step 5 is very viscous it is necessary to add TE(50/10), otherwise there will be incomplete separation of the phases after PCI extraction.

8   Any remaining traces of phenol are removed by an ether extraction. There is no need to centrifuge to separate the phases. Remove the upper (ether) phase by aspiration and repeat the extraction. Final traces of ether may be removed by blowing compressed air over the aqueous phase.

9   The next step is ethanol precipitation. Firstly add 1/10 vol 3 M Na acetate and then add 2.5 vols 95% EtOH.

10 To remove RNA the precipitate obtained in step 9 is washed with 70% EtOH, dried briefly *in vacuo*, dissolved in 10 ml TE(50/10) and incubated with RNase A at a concentration of 100 μg/ml for 1 h at 37 °C.

11 The RNase-treated sample is deproteinised by phenol extraction, and the DNA is precipitated with ethanol as described in step 9.

Bibliography

Süβmuth, R., Eberspächer, J., Haag, R. and Springer, W. (1987) *Biochemisch-mikrobiologisches Praktikum* (Stuttgart: Georg Thieme Verlag).

2.3.6   *Isolation of Total RNA from* Escherichia coli

The method chosen for the isolation of total RNA is a modified version of that described by Chirgwin *et al.* (1979). All manipulations carried out for isolation of RNA must be performed using sterile solutions and tubes etc. which have been sterilised by autoclaving. (Comments on RNAse contamination and the preparation of RNAse-free glass and plastic ware are to be found in Sambrook *et al.*, 1989.)

Material

Homogenisation buffer (4.5 M guanidium hydrochloride, 0.5 M Tris-HCl, pH 7.6, 2% Na-lauryl-sarcosinate); *phenol*, saturated with 0.1 M Tris-HCl, pH 6.5; *sodium acetate buffer* (100 mM Na acetate, pH 4.8); *PCI* (phenol, chloroform, isoamyl alcohol in the ratio 25 : 24 : 1 and saturated with TE(10/1); *CI* (chloroform/isoamyl alcohol in the ratio 24 : 1 and saturated with TE(10/1)); 3 M Na acetate; 5 M LiCl; EtOH; sterile tubes etc.; sterile H_2O; *E. coli* strains.

Method

1 100 ml LB broth (in a 1 l Erlenmeyer flask) is inoculated with 0.4 ml of an O/N culture of the bacterial strain and incubated, with shaking, at 37 °C until an OD_{600} of 1.0 has been reached.

2 After harvesting the cells by centrifugation for 10 min at 4 °C and 6000 rpm (JA-14 rotor) the cell pellet is resuspended in 5 ml homogenisation buffer. 5 ml phenol saturated with 0.1 M Tris-HCl, pH 6.5, is added and the suspension is incubated with gentle shaking at 65 °C. Following the addition of 2 ml 100 mM Na acetate, pH 4.8, the sample is incubated for a further 10 min at 65 °C (continue gentle shaking).

3 An equal volume of chloroform/isoamyl alcohol is added, the whole is mixed thoroughly and then centrifuged to separate the phases (JA-20, 6000 rpm, 10 min, 4 °C). The aqueous (upper) phase is transferred to a new tube and the organic phase is extracted a second time with phenol, shaking at 65 °C etc.

4 The combined aqueous phases are extracted with 10 ml TE(10/1)-saturated chloroform / isoamyl alcohol, centrifuged to separate the phases and the aqueous phase transferred to a new tube.

5 0.1 vol 3 M Na acetate and 2.5 vol EtOH are added to the aqueous phase, and the RNA is precipitated O/N at −20 °C.

6 The precipitated RNA is collected by centrifugation for 30 min at 6000 rpm and 4 °C in a JA-20 rotor. The RNA precipitate is dissolved in 2.5 ml sterile H_2O with gentle shaking in an ice-water bath. After the RNA has dissolved, 2.5 ml 5 M LiCl is added and the incubation in the ice-water bath is continued for a further 30 min.

7 The precipitated RNA is centrifuged, washed three times with 70% EtOH, dried *in vacuo* and dissolved in 1 ml sterile H_2O.

Determination of the Concentration and Purity of RNA

The concentration of RNA is measured photometrically. The purity of the RNA is determined from a UV spectrum. The quality of the RNA is determined by running a 1% agarose gel. The gel is photographed after staining with EtBr. In a good RNA preparation the 23S rRNA should fluoresce more strongly than the smaller 16S rRNA. The RNA can be transferred from the gel to a filter for Northern analysis.

References

CHIRGWIN, J.M., PRZYBYLA, A.E., MACDONALD, R.J. and RUTTER, W.J. (1979) Isolation of biologically active ribonucleic acid from sources enriched in ribonuclease. *Biochemistry* **18**, 5294–9.

SAMBROOK, J., FRITSCH, E.F. and MANIATIS, T. (1989) *Molecular cloning – a laboratory manual*, 2nd edn, Vol. II, section 7 (Cold Spring Harbor, NY: Cold Spring Harbor Laboratory Press).

2.3.7 Isolation of Chromosomal DNA from Saccharomyces cerevisiae

Material

YEP (20 g sucrose, 5 g yeast extract, 6.7 g peptone per litre dH$_2$O; solid medium contains in addition, 5 g KH$_2$PO$_4$, 5 g K$_2$HPO$_4$ and 20 g agar); *S. cerevisiae* strain e.g. X-2180-1A, ATCC 26786 (MATa SUC2 mal gal2 CUP1) [obtainable from the National Collection of Yeast Cultures (NCYC, Fax +44 1603 458414), Institute of Food Research, Norwich Research Park, Colney, Norwich NR4 7UA, UK or Yeast Genetic Stock Center, Department of Biophysics and Medical Physics, University of California, Berkeley, USA]; *SCE* (1M sorbitol, 0.1 M tri-sodium citrate, 10 mM EDTA, pH 5.8); *Zymolyase* (5 mg/ml); *SDS* (20% w/v); *NaCl/EDTA* (0.15 M NaCl, 0.1 M EDTA, pH 8.0); *proteinase K* (EC 3.4.21.4) (1 mg/ml in NaCl/EDTA, prepare immediately prior to use); *PCI* (phenol/chloroform/isoamyl alcohol; 25 : 24 : 1; phenol should be saturated with TE(10/1)); *CI* (chloroform/isoamyl alcohol, 24 : 1; TE(10/1); EtOH; *RNAse A* (EC 3.1.27.5), prepared as described in Subsection 2.3.5).

Method

1   5 ml YEP is inoculated with a single colony of the *S. cerevisiae* strain, and the culture is incubated for 20 h at 30 °C on a rolling drum.

2   The cells are harvested by centrifugation and placed at -70 °C for at least 2 h.

3   The frozen pellet is thawed and resuspended in 0.5 ml SCE, 10 μl Zymolyase is added and the suspension is incubated at 30 °C until about 90% of the cells have been converted into spheroplasts The spheroplast formation can be followed microscopically: 10 μl of cell suspension is placed on a microscope slide; while the cells are viewed in the microscope, 10 μl 10% SDS is introduced under the coverslip; the spheroplasts 'burst' in the presence of SDS.

4   The spheroplasts are harvested by centrifugation and the pellet is resuspended in 0.5 ml NaCl/EDTA. Following the addition of 100 μl of a freshly prepared solution of proteinase K and 50 μl 20% SDS the sample is incubated at 37 °C for 3–4 h and then at 60 °C for 20 min.

5   The sample is allowed to cool to room temperature before extracting with 1 vol PCI. The phases are separated by centrifugation.

6   The aqueous phase is transferred to a fresh reaction tube, and 2 vol EtOH (-20 °C) is added. Mix gently but thoroughly before centrifuging for 10 min. The white precipitate is dried briefly *in vacuo* and dissolved in 0.5 ml TE(10/1). 2 μl RNAse A (10 mg/ml) is added and the sample is incubated at 37 °C for 1 h. After extraction with 1 vol CI the DNA is precipitated with 2.5 vol EtOH, the precipitate dried *in vacuo* and allowed to dissolve O/N at room temperature in 50 μl TE(10/1).

Further reading

BICKNELL, J.N. and DOUGLAS, H.C. (1970) Nucleic acid homologies among species of *Saccharomyces*. *J. Bacteriol.* **101**, 505–12.

2.3.8   *Transformation of* Saccharomyces cerevisiae

Principle of the Experiment

In contrast to other eukaryotes the yeast *Saccharomyces cerevisiae* is amenable to both the traditional and 'modern' methods of genetics. As is the case with bacteria, yeast cells can be made 'competent', i.e. capable of taking up DNA. Removal or permeabilisation of the cell wall is a prerequisite for making yeast cells competent. Generally the cell walls are removed by treating the cells with enzymes having glucanase activity – there are several commercially available, e.g. Glusulase, Zymolyase. In the meantime this method has been adopted for fungi such as *Neurospora crassa* and *Aspergillus nidulans*.

Instead of making protoplasts, one can permeabilise *S. cerevisiae* cells by exposing them to high concentrations (1 M) of lithium ions. The technique described below uses this method. Regardless of which method one uses for permeabilising yeast cells the DNA to be transformed must fulfil certain requirements. The minimal requirement is that the uptake of the DNA can be selected for; therefore vectors carrying specific yeast 'marker' genes – e.g. the URA3 gene, coding for orotidine-5'-phosphate decarboxylase (EC 4.1.1.23) – are essential. There are four types of vectors known in yeast (for details see Kingsman and Kingsman, 1988). One of the differences between the vectors is the rate at which they transform yeast cells. In the experiment described below, the different transformation rates for each type of plasmid are determined.

Description of the Experiment

Transformation of a yeast strain carrying a mutation, ura3-52, in its URA3 gene with the plasmids YIp5 and YCp50 is described. The ura3-52 mutation is non-reverting, i.e. stable, caused by Ty insertion into the URA3 gene (Rose and Winston, 1984).

The properties of the two vectors are summarised in Table 2.2 and their structures are shown in Figs 2.3 and 2.4. In addition to the URA3 gene the plasmids contain an origin of replication (ori) allowing them to replicate in *E. coli* and, to permit selection in this bacterium, the genes encoding ampicillin (amp^r) and tetracycline

Table 2.2   Properties of the vectors YIp5 and YCp50.

Vector	Size (kb)	Markers			
		URA3	ARS1	2 μ	CEN
YIp5	5.5	+	–	–	–
YCp50	7.9	+	+	–	+

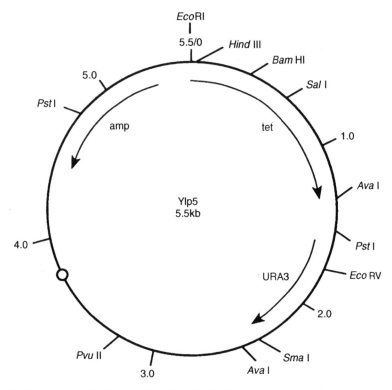

Figure 2.3 Structure of the *S. cerevisiae* plasmid YIp5.

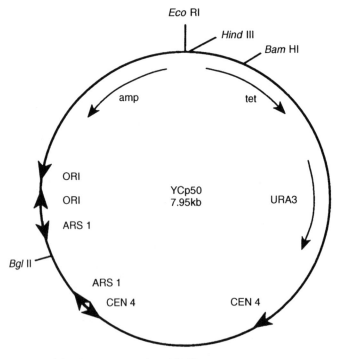

Figure 2.4 Structure of the *S. cerevisiae* plasmid YCp50.

(tet^r) resistance. However, the two plasmids differ in that one, YCp50 is capable of replicating in *S. cerevisiae* because it carries a yeast origin of replication, ARS1. This plasmid also contains another chromosomal element, CEN4, which is the centromere from chromosome IV of *S. cerevisiae* and, like all centromeres (one per chromosome), ensures the correct distribution of the chromosome at cell division.

Material

Yeast strain, e.g. DBY 747, ATCC 44774 (MAT a ura3-52 leu2-3 leu2-112 his3-1 trp4-289) (available from NCYC UK); *plasmids* YIp5 and YCp50 (1 mg/ml); *TE(10/1)*; *Li acetate (0.2 M in TE(10/1))*; *PEG 4000 (50%)* Plates: *SC-U (synthetic complete medium without uracil)* (Yeast Nitrogen Base w/o amino acids 0.67%, sucrose 2%, agar 2%; 20 μg/ml each of adenine sulphate, L-arginine-HCl, L-histidine-HCl, L-methionine, L-tryptophan; 30 μg/ml each of L-isoleucine, L-lysine-HCl, L-tyrosine; 50 μg/ml L-phenylalanine; 60 μg/ml L-leucine); *YEP plates* (see Subsection 2.3.7). Sterile test tubes (12 ml); sterile reaction tubes (1.5 ml); pipettes (glass and automatic); table top centrifuge; Eppendorf centrifuge (or similar); water bath with shaker.

Method

1 Inoculate 5 ml YEP medium with a single colony of the yeast strain to be used as the recipient in the transformation and incubate on a rolling drum for 24 h at 30 °C. Inoculate a predetermined amount (0.2 ml is roughly equivalent to 4×10^7 cells) of this O/N culture into 25 ml YEP medium in a 100 ml Erlenmeyer flask and shake at 30 °C until a concentration of $1-2 \times 10^7$ cells/ml has been reached. The volume of inoculum and incubation time required to reach the ideal cell density for transformation should be determined in a pilot experiment.

2 Transfer 10 ml aliquots to each of two sterile test tubes and harvest the cells by centrifugation at 5000 rpm for 3 min. Resuspend the cell pellets in the same volume of TE(10/1) and centrifuge again. The remainder of the log-phase culture is kept to determine the number of viable cells before treatment with Li$^+$ ions.

3 Resuspend each of the cell pellets from step 2 in 0.5 ml 0.2 M Li acetate in TE(10/1) and combine both suspensions in one tube. The cells are incubated in the presence of Li$^+$ ions at 30 °C with shaking for 60 min. *During this one-hour incubation set up the reaction tubes for the transformation according to the*

Table 2.3 Setting up the reaction tubes for yeast transformation.

Series No.	Control TE(10/1)	YIp5	YCp50
1	10 μl	10 μl	10 μl
2	20 μl	20 μl	20 μl

scheme shown in Table 2.3. Note that the transformation will be carried out with two different DNA concentrations; label plates and make the dilutions for determining the number of viable cells before lithium treatment. To determine viable cell count, make 10^{-4} and 10^{-5} dilutions of the log-phase cells in sterile H_2O. Plate 0.1 ml of each dilution onto each of two YEP plates and incubate them at 30 °C for 2–3 days.

4   Pipette 140 μl of the Li acetate suspension into each of the reaction tubes containing plasmid DNA or, for the control, TE(10/1). Mix briefly on the Whirlimix and leave the reaction tubes to stand at 30 °C for 30 min.

5   Add 350 μl 50% PEG 4000 to each reaction tube, mix thoroughly but carefully and allow the tubes to stand at 30 °C for a further 60 min.

6   Subject the samples to a 'heat shock' by placing them at 42 °C for 5 min and then allowing them to cool down to room temperature.

7   To reduce the viscosity of the samples add 500 μl sterile H_2O to each tube and mix.

8   Centrifuge the samples for 1 min at 2000 rpm (cells treated with Li^+ ions are very fragile!). Resuspend each cell pellet in 1 ml sterile H_2O by pipetting the suspension up and down a couple of times with the automatic pipette and centrifuge again at low speed. Wash the cells a second time with 1 ml H_2O and then resuspend the pellet in 100 μl H_2O.

9   Plate the control sample from the first series onto a SC-U plate. The number of colonies growing on this plate is a measure of the spontaneous reversion rate. The control of the second series is, on the other hand, diluted 10^{-4} and 10^{-5} in sterile dH_2O. 0.1 ml of each dilution is plated onto each of two YEP plates. The number of colonies growing on these plates after 2–3 days incubation at 30 °C can be used to calculate the survival rate following lithium treatment.

10 From each of the samples containing DNA make a 10^{-1} dilution (10 μl of cells + 90 μl H_2O) and plate this dilution and the remainder of the sample onto a SC-U plate.

11 Incubate all plates at 30 °C for 3–4 days.

12 Count the number of colonies on each plate and enter the figure in the appropriate column of Tables 2.4. and 2.5. Calculate the transformation frequency per μg DNA.

Table 2.4 Determination of the viable cell count before and after treatment with Li^+ ions.

		Number of cells	
Strain	Dilution	Before Li^+ treatment	After Li^+ treatment
DBY 747	10^{-5}		
	10^{-6}		

Table 2.5 Calculation of the transformation frequency for the plasmids YIp5 and YCp50.

Plasmid	DNA (μg)	No. of transformants on SC-U plates		No. of transformants per μg DNA
		10^{-1} dilution	Undiluted	
YIp5				
Series 1	10			
Series 2	20			
YCp50				
Series 1	10			
Series 2	20			

Bibliography

CASE, M.E., SCHWEIZER, M., KUSCHNER, S.R. and GILES, N.H. (1979) Efficient transformation of *Neurospora crassa* by utilizing hybrid plasmid DNA. *Proc. Natl. Acad. Sci. USA* **76**, 5259–63.

FINCHAM, J.R. (1989) Transformation in fungi. *Microbiol. Rev.* **53**, 148–70.

GUTHRIE, C. and FINK, G.R. (1991) Guide to yeast genetics and molecular biology. *Methods Enzymol.* **194**.

ITO, H., FUKUDA, Y., MURATA, K. and KIMURA, A. (1983) Transformation of intact yeast cells treated with alkali cations. *J. Bacteriol.* **153**, 163–8.

KINGSMAN, S.M. and KINGSMAN, A.J. (1988) *Genetic engineering – an introduction to gene analysis and exploitation in eucaryotes* (Oxford: Blackwell Scientific Publications).

ROSE, M. and WINSTON, F (1984) Identification of a Ty insertion within the coding sequence of the *S. cerevisiae* URA3 gene. *Mol. Gen. Genet.* **193**, 557–60.

TUITE, M.F. and OLIVER, S.G. (1991) *Yeast: exploitation and research* (New York: Plenum Press).

2.3.9   *Oligonucleotide-directed Mutagenesis (Kunkel Method)*

Principle of the Experiment

Oligonucleotides may be defined as short-chain polynucleotides. They are synthesised on a so-called DNA 'synthesiser' (e.g. Applied Biosystems DNA synthesiser type 381 A) and are used most frequently as primers for DNA sequencing, as 'mismatch' primers for *in vitro* mutagenesis and as 'linkers' when creating new restriction sites in DNA fragments, enabling them to be joined to each other. The 'classical' way of creating mutants is to expose the organism to a mutagen and select for the desired phenotype, which often involves screening very large numbers of cells. It is, however, possible to mutate specific bases in the DNA by *in vitro* mutagenesis and then test the effect of the mutation on the organism. Several methods are available for performing *in vitro* mutagenesis, and in this section the method developed by Kunkel (Kunkel *et al.*, 1987) will be described. The principle of the method is illustrated in Fig. 2.5. A set of oligonucleotide primers mutated at

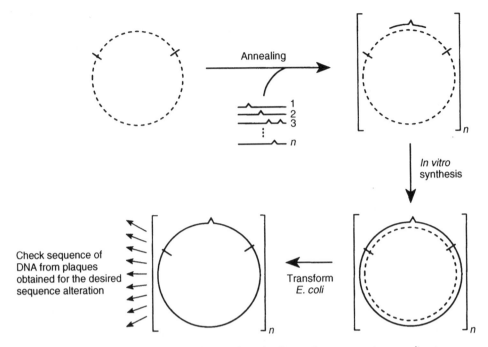

Figure 2.5   Schematic illustration of oligonucleotide-directed mutagenesis according to Kunkel.

different positions within the nucleotide sequence are annealed to the corresponding single-stranded DNA in which a certain proportion of the thymine residues are replaced by uracil and the complementary DNA strand is synthesised *in vitro* with T7 DNA polymerase. The reaction products are transformed into JM101, an *E. coli* strain which cannot maintain uracil-containing DNA, thus selecting for the mutated DNA. The individual mutants can be identified by sequencing.

Description of the Experiment

All *in vitro* mutagenesis methods use single-stranded DNA as the starting material, and this is hybridised to the mutated oligonucleotide. The nucleotide is complementary to the region of the sequence being investigated except for the base or bases it is intended to mutate. In this method the mutated DNA strand is selected for by degradation of the uracil-containing DNA in an appropriate *E. coli* strain. The single-stranded template DNA is synthesised in a strain of *E. coli* carrying the mutations dut⁻ and ung⁻ in the genes encoding the enzymes desoxyuridyltriphosphate hydrolase (dUTPase; EC 3.6.1.23) and Uracil-N-glycosylase, respectively. This combination of mutant enzymes means that any DNA synthesised in this strain contains a high proportion of uracil residues for the following reasons: (i) the intracellular concentration of dUTP increases because of the defective dUTPase, and dUTP can compete with dTTP for incorporation into the DNA; (ii) normally, dUTP residues are removed by the action of uracil-N-glycosylase, but since this enzyme is also defective the uracil residues remain in the DNA. Uracil-containing template DNA is essentially the same as wild type single-stranded DNA: both have the same coding capacity, and uracil is not a mutagen either *in vivo* or *in vitro*. The mutated oligonucleotide is annealed to the uracil-substituted template DNA, and a double-stranded molecule is synthesised. The products of this reaction are transfected into a dut$^+$ ung$^+$ strain. During phage multiplication the uracil-containing DNA is selectively degraded by uracil-N-glycosylase. This biological selection results in an enrichment of the DNA strand carrying the mutation(s). Using this method 60–70% of the plaques obtained contain DNA carrying the desired mutation.

Cloning of the DNA Fragment to be Mutagenised in the M13 Vector

Material

LB (see Subsection 2.3.1). *E. coli*: RZ1032[Hfr-KL16 Po/45{*lysA*(61-62)} *dut1 ung1 relA1 supE44 thi*]; JM101[Δ*(lac-proAB) supE thi* (F'*traD36 proAB$^+$ lacI^{q}Z* ΔM15)]. *PEG/NaCl* (PEG 6000 15% (w/v), NaCl 2.5 M); TE(10/1); *10 × T4 polynucleotide kinase buffer* (Tris-HCl 0.5 M, MgCl$_2$ 0.1 M, dithiothreitol 50 mM, spermidine 1 mM, EDTA 1 mM, pH 7.6); *ATP* 10 mM; *T4 polynucleotide kinase* (EC 2.7.1.78); *5 × sequencing buffer* (Tris-HCl 200 mM, MgCl$_2$ 100 mM, NaCl 250 mM, pH 7.5); *dNTP* (2 mM; 2 mM dATP, 2 mM dGTP, 2 mM dTTP, 2 mM dCTP); Sequenase [Sequenase from USB, USA; or T7 DNA polymerase (EC 2.7.7.7), e.g. Pharmacia]; T4 DNA ligase (EC 6.5.1.1); *PCI* (phenol (TE-saturated)/chloroform/isoamyl alcohol in the proportion 25 : 24 : 1); *CI* (chloroform/isoamyl alcohol in the proportion 24 : 1); siliconised glass wool; reaction tubes (0.5 ml and 1.5 ml); *Na acetate* 3 M; EtOH; sterile centrifuge tubes; heating block; water baths, etc.

Example

The 0.65 kb *Eco*RI/*Sal*I fragment from pBR322 is to be cloned into the M13 mp18 vector cut with *Eco*RI and *Sal*I. Specifically, pBR322 is cut with *Eco*RI and *Sal*I, and the sample is subjected to electrophoresis in 1.2% agarose in 1 × TBE; two fragments with lengths of 3.7 and 0.65 kb are obtained. The 0.65 kb fragment is isolated from the gel by the 'freeze–squeeze' method and ligated into *Eco*RI/*Sal*I-restricted M13mp18 vector DNA and transformed into *E. coli*. Double-stranded DNA is isolated from a number of white plaques and examined by restriction analysis. The phage titre of a plaque containing the correct insert is determined. The final step is the isolation of the uracil-substituted single-stranded DNA from the phage plaque. The purpose of this technique is to mutate the *Bam*HI restriction site, located within the *Eco*RI/*Sal*I restriction fragment, from 'GGATCC' to 'GGAACC'. As a result of this mutation the *Eco*RI/*Sal*I fragment can no longer be cut by *Bam*HI.

Isolation of DNA Fragments from Agarose Gels by the 'Freeze–Squeeze' Method

Normally, DNA fragments are isolated from preparative gels, but acceptable results can be obtained using analytical gels.

Method

1 At the end of the electrophoresis run, the DNA fragments are visualised under

UV light and the fragment to be isolated is cut out of the gel with a flamed scalpel or razor blade, avoiding 'excess' agarose.

2 The agarose strip is placed in a 0.5 ml reaction tube containing siliconised glass wool and perforated 5 or 6 times close to the base. This tube is then placed inside a 1.5 ml reaction tube and both are placed at $-70\,°C$ for at least 4 h.

3 The 'tube within a tube' is centrifuged for 10 min at room temperature, causing fluid from the agarose and DNA to be forced via the perforations from the smaller tube into the larger. Agarose particles are retained by the glass wool.

4 1/10 vol 3 M Na acetate and 2.5 vol EtOH are added to the DNA-containing fluid and the whole is placed at $-70\,°C$ for 2 h. The precipitated DNA is collected by centrifugation. After drying *in vacuo*, the DNA is dissolved in $10-20\ \mu l$ TE(50/10).

Cloning the 0.65 kb *Eco*RI/*Sal*I Restriction Fragment from pBR322 in M13 mp18 Vector DNA Cut with the Same Restriction Enzymes

Method

1 Cut M13 mp18 DNA with *Eco*RI and *Sal*I.

2 Ligate the 0.65 kb *Eco*RI/*Sal*I fragment from pBR322 with the appropriately restricted vector DNA from step 1.

3 Transform aliquots of the ligation sample into *E. coli* (e.g. JM101); add X-Gal and IPTG before plating out the cells (see Subsection 2.3.4).

4 'Blue/white' selection.

Isolation of Double-stranded M13 Recombinant DNA

Method

1 Using sterile yellow tips or pasteur pipettes transfer 'white' plaques singly into 2.5 ml $2 \times$ TY broth (or LB broth) which has been inoculated with $20\ \mu l$ of a fully grown culture of the *E. coli* strain JM101. Incubate the tubes at $37\,°C$ for at least 6 h.

2 Decant the cultures into reaction tubes and centrifuge for 30 s. Carefully transfer the supernatant, which contains phage particles, to sterile reaction tubes and store at $4\,°C$. Isolate double-stranded DNA from the cell pellet by the 'mini-screen' method.

3 Analyse DNA by restriction digest and electrophoresis.

Titration of M13 Phage Suspension

Method

1 Inoculate 10 ml $2 \times$ TY broth (or LB broth) with $50\ \mu l$ of an O/N culture of JM101 and a 'white' plaque taken from the plate as described in step 1 above. Incubate the culture, with shaking, for 6–7 h at $37\,°C$.

2   Centrifuge the culture (JA-20 rotor) for 10 min at 6000 rpm and repeat the centrifugation step. Store the supernatant, containing the phage particles, at 4 °C.

3   Dilute the phage suspension to 10^{-6} to 10^{-8} in TE(10/1). Mix 100 μl of each dilution with 100 μl of JM101 with 3 ml molten H top agar and pour onto prewarmed H plates.

4   The plaques are counted after 12–16 h incubation at 37 °C. DO NOT FORGET THE DILUTION FACTOR!

Growth of Phage with Uracil-containing DNA

Method

1   20 ml 2 × TY or LB broth is inoculated with 50 μl of an O/N culture of the strain RZ1032 and incubated, with shaking, at 37 °C for 4 h.

2   10^{10} to 10^{11} phage particles (usually equivalent to 50–200 μl) are added and the incubation is continued for a further 5 h.

3   The cells are harvested (as described above) and the supernatant is used to inoculate a second culture of RZ1032. The titre of the phage-containing supernatant is determined as described above but using RZ1032 as the bacterial lawn.

4   This process is repeated twice. The phage-containing supernatant from the fourth passage through RZ1032 is used to isolate the single-stranded template DNA to be used in the mutagenesis.

Preparation of Single-stranded M13 Recombinant DNA

Method

1   The volume of the phage-containing supernatant is determined. 0.25 vol PEG/NaCl is added and the tube contents mixed by inverting the tube several times. The sample is placed on ice for 1 h.

2   The precipitated phage particles can be collected by centrifugation (JA 20, 30 min at 4 °C and 900 rpm); the PEG-containing supernatant is removed by aspiration with a pasteur pipette connected to a vacuum pump via rubber tubing. The tube is inverted for a few seconds on a paper towel to remove any remaining traces of PEG/NaCl.

3   The phage pellet is resuspended in 4 ml TE(10/1) and transferred to a 15 ml Corex tube. The tube is rinsed with 2 ml TE(10/1), and this is added to the suspension in the Corex tube.

4   The suspension is mixed vigorously for 30 s on a Whirlimix and centrifuge.

5   The supernatant is carefully transferred to a new tube, avoiding as far as possible taking any of the pellet which contains bacterial debris. The supernatant is extracted twice with an equal volume of PCI and then once with CI. For each extraction phase, separation is achieved by centrifuging for 5 min at 10 000 rpm and 20 °C (JA-20 rotor).

6   0.1 vol 3 M Na acetate is added to the final aqueous phase before adding 1 vol EtOH. The sample is placed at $-70\,^{\circ}C$ for 1–2 h. The precipitated DNA is collected by a 10-minute centrifugation, washed with 70% EtOH, dried *in vacuo* and dissolved in 200 μl TE(10/1).

7   The concentration of the DNA is determined by measuring OD_{260} (1 $OD_{260} = 36$ μg/ml), and an aliquot is analysed by gel electrophoresis and compared in size with the original single-stranded DNA.

Phosphorylation of the Mutated Oligonucleotide at its 5'-OH End

The oligonucleotide must be phosphorylated at its 5'-OH end before the DNA polymerase reaction is carried out, otherwise it cannot be ligated to the 3'-OH end of the newly-synthesised DNA.

Method

- 100 pmol mutated oligonucleotide (see discussion of sequence, below)
- 2 μl 10 × T4 polynucleotide kinase buffer
- 1 μl 10 mM ATP
- 4 U T4 polynucleotide kinase.

Make up to 20 μl with ddH$_2$O and incubate for 1 h at $37\,^{\circ}C$. Hold the sample at $68\,^{\circ}C$ for 10 min to inactivate the kinase.

pmol DNA $= 1 \times 10^6/(660 \times$ No. of bases)
660 g/mol = average molecular weight of one bp (Na salt).

The difference between the wild type and the mutated oligonucleotide is shown by the lower case letter in the sequence below.

5'-GTCCTGT**GGATCC**TCTACGCCG-3'
5'-GTCCTGTGGAaCCTCTACGCCG-3'

The sequence of the oligonucleotide used for the mutagenesis corresponds to nucleotides 367 to 389 in the nucleotide sequence of pBR322 (Sutcliffe, 1979). The *Bam*HI restriction site in the wild type sequence is shown in bold type. The oligonucleotide should be chosen such that the sequences flanking the mutated base are 8–10 nucleotides long.

Mutagenesis

Method

1   Hybridising the 'primer' (mutated oligonucleotide) to the template DNA:
   (i)   1.5 μl (0.2 pmol) M13 single-stranded DNA
   (ii) 1.0 μl (1 pmol) 5'-phosphorylated oligonucleotide
   (iii) 2.0 μl 5 × sequencing buffer
   (iv) 5.5 μl H$_2$O.

The sample is mixed and incubated at 65 °C for 5 min either in a heating block or in a water bath. After this short incubation at the higher temperature the sample is cooled down to 30 °C over a period of at least 20 min. The following empirical formulae allow the calculation of the optimal temperature for hybridisation of oligonucleotides:

$$T_m = 2\,°C \times (A + T) + 4\,°C \times (G + C)$$
$$T_e = T_m - 15\,°C$$

where T_m is melting temperature, T_e is temperature of the experiment.

2 The following solutions are now added to the sample:

(i)   7 μl 5 × sequencing buffer

(ii) 6 μl dNTP stock solution (each dNTP at 2 mM)

(iii) 4.5 μl 10 mM ATP

(iv) 11.5 μl H_2O

(v)   1 μl Sequenase (12.5 U)

(vi) 5 μl T4 DNA ligase (5 U).

3 The sample is placed on ice for 5 min, then 5 min at room temperature and finally it is incubated at 37 °C for 15 min. The enzyme is inactivated by heating the sample to 65 °C for 5 min. The lower temperature favours the initiation of DNA synthesis from the 3'-OH group of the oligonucleotide, whereas at 37 °C the efficiency of the elongation reaction is increased.

4 One-half of the sample is transformed into JM101 (to enrich for the mutated strand), and the double-stranded DNA is isolated from single plaques. For control purposes the rest of the sample is transformed into RZ 1032.

5 The presence of mutated DNA can be checked by restriction analysis: the 0.65 kb *Eco*RI/*Sal*I fragment cannot be cut with *Bam*HI. For control purposes the wild type fragment digested with *Bam*HI should also be loaded onto the gel.

References

HORWITZ, B.H. and DIMAIO, D. (1990) Saturation mutagenesis using mixed oligonucleotides and M13 templates containing uracil. *Meth. Enzymol.* **185**, 599–611.

KUNKEL, T.A., ROBERTS, J.D. and ZAKOUR, R.A. (1987) Rapid and efficient site-specific mutagenesis without phenotypic selection. *Meth. Enzymol.* **154**, 367–82.

SUTCLIFFE, J.G. (1979) *Cold Spring Harbor Symposia* **43**, 77–83 (Cold Spring Harbor, NY: Cold Spring Harbor Laboratory Press).

2.3.10 Southern Transfer of DNA from Agarose Gels onto Nitrocellulose or Nylon Membrane

Principle of the Experiment

Southern transfer or blotting is the technique by which DNA, after electrophoresis through an agarose gel, is transferred from the gel onto a carrier membrane and 'fixed' to the membrane. The method was developed by Southern in 1975 while he was at the University of Edinburgh. Other 'blotting' techniques, e.g. the transfer of RNA or protein from gels to a carrier membrane, are known as 'northern' and 'western' blotting, respectively, in acknowledgement of Southern's pioneering work.

Description of the Experiment

The experimental set-up for Southern blotting is illustrated in Fig. 2.6.

Material

Denaturing solution: (NaOH 0.5 M, NaCl 1.5 M); *neutralising solution* (Tris-HCl 0.5 M, NaCl 1.5 M, pH 7.5); *20 × SSPE* (NaCl 3.6 M, EDTA 0.02 M, NaH_2PO_4 0.2 M, pH 7.7); Whatman 3 MM paper; nitrocellulose (0.45 μm pore size) or nylon membrane (e.g. Hybond N-filter from Amersham); container; glass plate, etc.

Method

1   The agarose gel containing the electrophoretically separated DNA (stained with EtBr and photographed) is immersed in denaturing solution for 15 min (gentle shaking is possible, though not essential). The solution is discarded and the procedure repeated. The gel is rinsed briefly with H_2O before immersing in neutralising solution for 15 min. Finally the gel is shaken gently in 2 × SSPE for 3 min.

2   The Southern transfer is set up as follows: a shallow container is filled with 20 × SSPE, and over this is placed a glass plate. A strip of Whatman 3 MM paper is cut so that the width is the same as the length of the gel and the length is sufficient for it to hang over the glass plate into the 20 × SSPE.

3   Three sheets of Whatman 3 MM paper, cut to the size of the gel and wetted with 20 × SSPE, are placed on the paper, acting as a 'wick' and on top of this is placed the gel followed by the nitrocellulose or nylon membrane cut to the size of

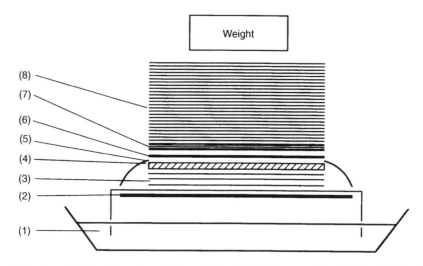

Figure 2.6   Cross-section through a Southern blot: (1) shallow container filled with 20 × SSPE; (2) glass plate supporting the strip of Whatman 3MM paper soaked in 20 × SSPE which dips into the reservoir of 20 × SSPE; (3) three sheets of Whatman 3MM paper, soaked in 20 × SSPE; (4) agarose gel; (5) Parafilm™ or cling-film to prevent the gel drying out; (6) nitrocellulose or nylon membrane, soaked in 2 × SSPE; (7) three sheets of Whatman 3MM, soaked in 2 × SSPE; (8) stack of absorbent paper.

the gel and moistened with $2 \times$ SSPE. Three 'gel size' sheets of Whatman 3 MM, wetted with $2 \times$ SSPE, are placed onto the nitrocellulose or nylon membrane followed by two dry sheets of Whatman 3 MM, again cut to gel size. It is important that no air bubbles form between the layers; air bubbles can be eliminated by rolling a 10 ml glass pipette over each layer before adding the next. Finally, a stack of paper hand towels, cut to gel size and 5–10 cm high, is placed on top of the filter papers. A glass plate and a weight of approximately 500 g is placed on top to complete the set-up. To prevent a 'short circuit' of the transfer, Parafilm^{TM} or cling-film is placed around the gel.

4 The length of time required for transfer lies between 5 and 16 h.

5 After dismantling the transfer the membrane filter is placed in $2 \times$ SSPE for 2–3 min and, in the case of nitrocellulose, baked *in vacuo* for 1–1.5 h at 80 °C. For nylon membranes follow the instructions of the manufacturer.

Reference

SOUTHERN, E.M. (1975) Detection of specific sequences among DNA fragments separated by gel electrophoresis. *J. Mol. Biol.* **98**, 503–17.

2.3.11 Detection of Nucleic Acids by Non-radioactive Methods

Principle of the Experiment

After the DNA has been transferred to the nitrocellulose or nylon membrane it can be detected by hybridisation with a radioactive or non-radioactive labelled probe. The techniques using non-radioactive labelling of DNA are becoming more and more popular, not least because of the risks involved in working with radioactive isotopes. Another advantage of non-radioactive labelling is the increased stability of the labelled probes compared with those obtained with radioactive isotopes. Kits are available from several companies for detection of nucleic acids with non-radioactive methods. Labelling with *biotin* is one such method in which biotinylated desoxynucleotides are incorporated into the DNA. The next step is the hybridisation of the biotinylated DNA probe with the DNA bound to the filter and detection of the hybridisation complex with streptavidin and alkaline phosphatase. Streptavidin can bind to the biotinylated DNA as well as to the biotin-labelled alkaline phosphatase since it has four binding sites for biotin. Following the addition of a chromogenic substrate a violet-coloured precipitate forms. The intensity of the colour is increased because of the complexing of several streptavidin/biotin-labelled enzyme molecules.

The detection of membrane-bound nucleic acids is also possible by means of chemiluminescence, when the substrate molecule is dephosphorylated by alkaline phosphatase and light is emitted.

Yet another method, which involves labelling the DNA with *digoxigenin*, has been developed by Boehringer. Digoxigenin is a steroid derivative isolated from digitalis plants. In this method the DNA is labelled with the nucleotide analogue digoxigenin-11-dUTP and hybridisation of the probe to membrane-bound DNA is detected by an enzyme-coupled reaction to an antibody which recognises digoxigenin.

The 'enhanced chemiluminescence' system from Amersham is another non-radioactive detection method which can be highly recommended.

Description of the Experiment

Each company offering a non-radioactive detection system provides detailed information relevant to the particular method. For this reason no further information will be given here.

Further reading

Information about ECL (Amersham).
Non-radioactive labelling and detection: an application manual (Boehringer Mannheim Biochemica).
Gene images – non-isotopic nucleic acid labelling and detection systems (Cleveland, Ohio: United States Biochemicals (USB)).

2.3.12 In Vitro *Synthesis of Specific DNA Fragments with the Polymerase Chain Reaction*

Principle of the Experiment

A specific DNA fragment can be amplified by means of the polymerase chain reaction (PCR). PCR is an *in vitro* method for the enzymatic synthesis of specific DNA sequences starting from two strand-specific oligonucleotide primers annealed to the opposite ends of the template DNA. A series of repetitive cycles consisting of denaturing the template DNA, hybridisation of the primers to the denatured template and extension of the primer on the template to yield a double-stranded molecule leads to an exponential increase in the number of DNA molecules. The reaction products of one cycle serve as templates for the next cycle, thus doubling the number of molecules every cycle. In 20 PCR cycles, the molecule being amplified has been produced 1×10^6 times.

The method, developed originally by Mullis and Faloona (1987), is now used routinely in the laboratory to check whether or not cloning steps have been carried out correctly. The DNA is synthesised using the heat-stable DNA polymerase from the thermophilic bacterium *Thermus aquaticus* (Taq polymerase; EC 2.7.7.7). This enzyme has optimal activity at approximately 70 °C and is stable at 95 °C during the denaturation phase of each cycle. The advantage of Taq polymerase in comparison with the heat-labile T7 DNA polymerase or Klenow enzyme is that the process lends itself to automation because there is no need to add a fresh aliquot of enzyme to the sample at the end of each cycle.

PCR has become a universal technique for DNA detection and is used, for example, for producing phylogenetic family trees or for the prenatal diagnosis of sickle cell anaemia.

Description of the Experiment

The conditions for performing PCR cycles vary according to template DNA, primer used, hybridisation temperature, concentration of Mg^{2+}, etc. Therefore the exact conditions have to be determined empirically for each template DNA. This is particularly important for the temperature at which the primer is annealed to the template, since the dissociation temperature for an A–T pair is lower than that for a G-C pair. It is recommended that the annealing temperature is about 4 °C lower than the dissociation temperature.

The products of the PCR reaction have to be purified prior to cloning. The cloning of PCR products can be a problem because the ends of the product may be 'ragged', i.e. heterogeneous. One method is to 'fill in' the ends of the PCR product with T4 DNA polymerase. Another possibility is to design the primers so that they carry restriction sites 2–3 bases from their ends. Following a restriction digest the PCR product can be cloned into the appropriately restricted vector. Frequently, two different restriction sites are used, a strategy known as 'forced cloning'.

The 'TA cloning' method developed by Mead *et al.* (1991) exploits the fact that Taq polymerase catalyses the template-independent addition of a single nucleotide, *viz.* deoxyadenosine, at the 3' ends of the PCR products. The commercially available linearised vector containing single-stranded 3'-T overhangs within the polylinker can be used to clone the PCR product because of its overhanging A nucleotides. This vector is available from Invitrogen.

In Vitro Amplification of Double-stranded DNA

Material

10 × Taq polymerase buffer; *dNTP mix* (2 mM dATP, 2 mM dGTP, 2 mM dTTP, 2 mM dCTP); *Taq polymerase* (EC 2.7.7.7); *TE(10/1)*; *Na acetate* 3 M; *EtOH*; sterile tubes; *DNA templates*; *primers*; PCR apparatus (e.g. Hybaid or Thermocycler 60 from Biomed); *paraffin oil*.

Method

1 The following are pipetted into a 0.5 ml reaction tube:
   (i)   100 ng plasmid DNA
   (ii)   100 ng of each primer
   (iii) 10 μl 10 × Taq polymerase buffer
   (iv)   10 μl dNTP mix
   (v)   H$_2$O ad 99 μl
   (vi)   1 μl Taq polymerase diluted to 2 U/μl.

2 The sample is mixed thoroughly and overlaid with 100 μl paraffin oil. This prevents loss of the sample by evaporation.

3 The sample is placed in the thermocycler and heated to 93 °C for 90 s, thus denaturing the template; it is then held at 45 °C for 50 s to allow annealing of the primer onto the template; the temperature is now increased to 72 °C for 90 s and the elongation of the primer can take place. The next step is denaturation at 93 °C. The apparatus is programmed so that these steps are repeated 44 times; at each cycle the length of the time allowed for the elongation step is extended by 4 s to compensate for any loss in the activity of the enzyme. A second sample which differs only in the amount of template DNA (500 ng) added can be amplified at the same time.

4   At the end of the amplification cycle the samples are placed at −70 °C for 30 min. Upon thawing, the paraffin can be removed easily because it thaws before the aqueous phase does.

5   To test the success of the amplification, 8 μl of each sample is subjected to electrophoresis. Comparison with a standard confirms that the fragment has the expected length. The yield of fragment is approximately 10 μg.

Gel Purification of a PCR Product Containing a Restriction Site at Each End

Before ligation into a vector the PCR product must be freed from any remaining nucleotides, oligonucleotides, template DNA and any DNA fragments arising in the rare event of the unspecific annealing of the primers on the template.

Method

1   Add 1/10 vol 3 M Na acetate, pH 4.8 and 2.5 vol EtOH to 30 μl (approx. 3 μg) of the first sample, mix and place at −70 °C for 30 min. Centrifuge for 20 min at 16 000 rpm and 4 °C (JA-20 rotor with adapters for Eppendorf tubes). Wash the pellet with 70% EtOH and dry *in vacuo*.

2   Dissolve the DNA in 20 μl TE(10/1). Once the DNA has dissolved, restrict it with the appropriate enzymes in a volume of 40 μl for at least 2 h.

3   Following the addition of 4 μl of loading dye and incubation at 68 °C for 10 min, load the sample onto a 1.5% agarose gel and subject to electrophoresis.

4   Stain the gel with EtBr, cut out the DNA-containing bands and extract the DNA by the freeze–squeeze method (Subsection 2.3.9).

The DNA fragment can now be cloned into the chosen vector.

Further reading

EHRLICH, H.A. (1989) *Principles and applications of DNA amplification* (New York: Stockton Press).

MCPHERSON, M.J., QUIRK, P. and TAYLOR, G.R. (1991) *Practising PCR – a practical approach* (Oxford: IRL Press).

MEAD, D.A., PEY, N.K., HERRNSTADT, C., MARCIL, R.A. and SMITH, L.M. (1991) A universal method for the direct cloning of PCR amplified nucleic acid. *BIO/Technology* **9**, 657–63.

MULLIS, K.B. and FALOONA, F. (1987) Specific synthesis of DNA *in vitro* via a polymerase-catalyzed chain reaction. *Meth. Enzymol.* **155**, 335–50.

3

Bioreactor Technology

3.1 BASIC METHODS FOR DESIGNING BIOREACTORS

P. Götz

3.1.1 *Determination of Characteristic Parameters for the Design of Stirred Vessels*

Principles and Aims

The most frequently used reactor in biotechnology is a mixing vessel. In this type of reactor the contents are agitated, resulting in an increased surface area in aerated aerobic cultures and an improved temperature transfer (either heating or cooling) at the inner surface of the reactor. Depending on the requirements of agitation, different parameters must be known when designing and scaling-up propellers. To calculate the energy required, a dimensional analysis with the determinant values of the agitation process can be performed. A further important relationship serves to calculate the shear forces created by the propeller in the reactor. A dimensional analysis of the agitation process gives not only several geometrical relationships but also three terms. These terms serve as the dimensionless description of the power characteristics of a propeller. The Newton number, Ne, is a function of: N, the energy transferred to the liquid from the propeller; n, the rotational speed of the propeller; d_R, the diameter of the propeller; and ρ the density of the liquid stirred in the fermenter.

$$Ne = \frac{N}{n^3 d_R^5 \rho} \tag{3.1}$$

The Reynolds number, Re, is a function of n, d_R, ρ and the viscosity, η.

$$Re = \frac{n d_R^2 \rho}{\eta} \tag{3.2}$$

The energy transfer from the propeller to the fluid depends on the torque (M_d) and the speed of the propeller:

$$N = M_d \, 2\pi \, n \tag{3.3}$$

The influence of the third number, the Froude number, *Fr*,

$$Fr = \frac{n^2 d_R}{g} \tag{3.4}$$

where *g* is the acceleration due to gravity, is relevant only when $Re > 300$ and for the purposes here may be disregarded. Taking the fixed geometrical relationships into consideration the Newton or power number, *Ne*, may be considered to be the friction factor of the propeller. Observations with different flowpaths (tube, ball) show that, in general for fluid flow in the laminar part, the relation given in Eqn (3.5) can be expected.

$$Ne = c \, Re^{-1} \tag{3.5}$$

where *c* is the propellor constant. For increasing values of Reynolds number there is a turbulent region following the transition and linear regions in which the power number remains constant (Fig. 3.1). If the stirrer constant *c* is known, the stirrer can be used for viscosity measurements. For non-Newtonian fluids Reynolds number can be used as the term for the apparent viscosity η_a, giving Eqn (3.6).

$$Ne = \frac{N}{n^3 d_R^5 \rho} = \frac{c}{Re} = \frac{c\eta_a}{n d_R^2 \rho} \tag{3.6}$$

Eqn (3.6) in terms of η_a gives Eqn (3.7).

$$\eta_a = \frac{N}{n^2 d_R^3 c} \tag{3.7}$$

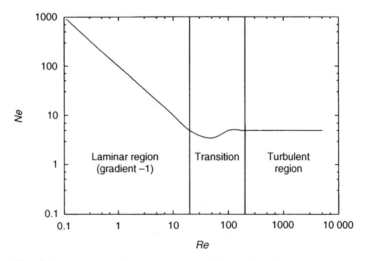

Figure 3.1 Plot of the power number against Reynolds number for different forms of fluid flow on immersed solid bodies.

Eqns (3.3) and (3.7) give Eqn (3.8).

$$\eta_a = \frac{2\pi M_d}{n d_R^3 c} \tag{3.8}$$

In the second part of the experiment described below, the relation between the rotational speed of the propeller and the shear forces produced will be determined using the above equations. This relation must be known for the coefficient of material transport in non-Newtonian fluids or for the prevention of damage to microorganisms by shear forces produced by agitation.

Description of the Experiment

Principle

The power of a propeller can be determined in the laminar and in the transition region by using a high-viscosity Newtonian fluid. The method developed by Metzner will be used to determine the shear rate. The principle of the method is that 'like causes like'. In this particular case the apparent viscosity of a non-Newtonian fluid may be caused by either the shear rate as defined in the annular clearance of a rotation viscometer or by the rotation speed of the propeller being used. This non-Newtonian fluid should obey the rule of Ostwald and de Waele to simplify the calculation. The viscosity of such a fluid can be determined from Eqn (3.9).

$$\eta_a = K_c D^{m-1} \tag{3.9}$$

where K_c is the index of consistency, D is the shear rate and m is the fluid index.

Equipment

VISCOSIMETER

A rotating viscosimeter with a thermostatically controlled sampling vessel is required. It should be possible to select the different rotational speeds manually and there should be a gauge showing the torque. Furthermore, several constants of the apparatus must be known, e.g. in order to measure the rate of shear in the annular opening. For measurements on model stirrers (e.g. disk propeller with $d_R = 0.1$ m) there must be the possibility of fixing the measuring device for the torque of the propeller or there should be an extra device for measuring the torque of the propeller. For viscosity measurements on Newtonian fluids a ball and cylinder viscosimeter can be used, but this is not essential for the experiment.

FLUIDS

Glycerol (>95% pure) can be used as a Newtonian fluid with a high viscosity, and a 5% aqueous solution of xanthane can serve as a non-Newtonian fluid.

Procedure

The first part deals with the power characteristic, $Ne = f(Re)$ for a turbine and a helical paddle and the graphical determination of the propeller constant c (in the laminar region) and Re_{crit}. The measurements for determining $Ne = f(Re)$ are performed with a Newtonian fluid. The specific gravity of the fluid is determined by volume measurement and weighing. The viscosity of the fluid can be determined using a ball and cylinder viscosimeter. The propeller, fixed to a torque-measuring device, is placed in a container filled with a Newtonian fluid, and the torque necessary for the energy transfer at various rotational speeds is noted. The geometry of the experimental set-up must be noted to allow comparison with other results. A sufficient number of measurements must be made in the laminar region, and the levelling-off of the curve in the transition region must be clearly visible. To determine the shear rate of the propeller using the method developed by Metzner, a non-Newtonian fluid should be used. The fluid is stirred at various speeds with a propeller attached to a torque-measuring device. At specified rotational speeds the corresponding torque is noted and the apparent viscosity calculated according to Eqn (3.8) using the stirring constants determined in the first part of the experiment. For the same fluid the relationship between shear rate and apparent viscosity is determined using a rotation viscosimeter. It must be ensured that the measurements are made in the laminar region and that the range of viscosity measurements for stirrer and annular clearance overlap (Fig. 3.2).

Interpretation, Conclusions and Commentary

For the first part of the experiment the experimental values are plotted on double logarithmic paper as shown in Fig. 3.2. The results of this graphical

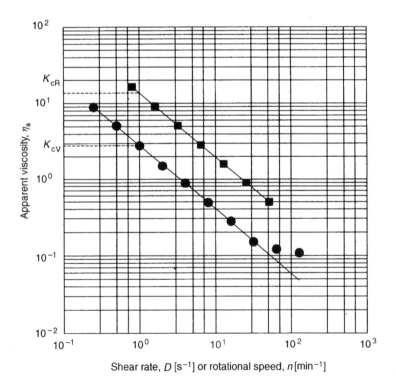

Figure 3.2 Determination of rotational speed, n, and rate of shear, D, as a function of apparent viscosity, η_a, of a fluid: ● viscosimeter; ■ stirrer.

interpretation are:

- limitation of the laminar region by Re_{crit}
- determination of the stirring constant c by reading Ne for $Re = 1$.

Both these values are necessary for carrying out the second part of the experiment.

For the interpretation of the second part of the experiment the viscosity values obtained with the stirrer as a function of the rotation of the propeller and with the rotational viscosimeter as a function of the shear rate are plotted on double log graph paper (Fig. 3.2). When the propeller measuring system is used there is a connection between rotation and apparent viscosity as described in Eqn (3.9), and it follows that:

$$\eta_a = K_{cR} n^{m-1} \tag{3.10}$$

This is equivalent to the assumption that the stirrer in the fluid causes an average shear gradient which is responsible for the observed apparent viscosity. The fluid index m in Eqns (3.9) and (3.10) must be independent of the measuring system, therefore the curves plotted must have the same slope. Equating Eqns (3.9) and (3.10) (subscript R stirrer, V viscosimeter) results in Eqn (3.11).

$$K_{cV} D^{m-1} = K_{cR} n^{m-1} \tag{3.11}$$

Expressing the equation in terms of D gives the relationship between rotational speed of the stirrer and the relevant average shear rate at the stirrer:

$$D = \left(\frac{K_{cR}}{K_{cV}}\right)^{\frac{1}{m-1}} n \tag{3.12}$$

The proportionality constant connecting rotation and shear rate at the stirrer is termed the Metzner constant, k.

It should be noted that Eqn. (3.12) is only valid for the region of laminar flow. When calculating the power of the propeller in the region of turbulent flow ($Ne = $ constant) the apparent viscosity may be disregarded since the power is independent of the viscosity. However, for the correlation of the transfer constant for oxygen, $k_L a$, the apparent viscosity in the region of turbulent flow is extremely important even when the culture is stirred. Determination of this constant can also be made with the help of Eqn. (3.12) for the region of linear flow if no other information is available.

3.1.2 *Rheology of Fermentation Fluids*

Principles and Aims

Frictional forces arise from the flow of real fluids. This internal friction is caused by molecular impulses within the fluid (exchange of impulses in a gas, cohesion in a fluid). In analogy to the laws applying to molecular and heat transport (Fick's first law and Fourier's first law) Newton's law of displacement links cause and effect, in this instance the mass of the impulse flow with the rate of velocity. The mass of the impulse flow is generally referred to as displacement, is designated by τ and is linked to the velocity gradient, D (also known as shear gradient) by the coefficient of dynamic viscosity, η.

$$\tau = -\eta \frac{dw}{dx} = -\eta D \tag{3.13}$$

For Newtonian fluids this viscosity is a constant dependent on pressure and temperature ($[\eta] = $ Pa s; η (water, 20 °C) = 1 mPa s).

By analogy with the movement of particles, energy and impulse, kinematic velocity v ($[v] = $ m^2/s) is also used to describe diffusion. The fluids in many biotechnologically important processes have non-Newtonian flow properties: i.e. the viscosity is not constant but is a function of the velocity gradient. In fermentation fluids the microorganisms (e.g. filamentous mycelia), the substrate (e.g. particulate matter) or the product (e.g. polysaccharides) can also contribute to their non-Newtonian properties. In spite of multiple flow behaviour properties the majority of fermentation fluids encountered in practice may be classified as purely viscous (see Table 3.1).

The purely viscous fluids can be divided into three groups on the basis of their flow properties:

1 pseudoplastic flow behaviour

2 dilatant flow behaviour

3 plastic flow behaviour (flow boundary).

Dilatant or pseudoplastic behaviour can be described mathematically (Eqn (3.14)) using the power term of Ostwald and de Waele (for the sake of simplicity Newton's law of displacement, which takes into account the vectorial nature of the quantities, will not be included below).

$$\tau = K_c D^m \tag{3.14}$$

where K_c is the index of consistency and m the flow index ($m < 1$ for pseudoplastic, $m = 1$ for Newtonian fluids and $m > 1$ for dilatants). The existence of a flow boundary is characteristic of plastic flow behaviour, i.e. the fluid starts to move only above minimum displacement. The simplest example of such substances is provided by Bingham fluids: once the minimum displacement τ_0 has been exceeded they start to flow and exhibit Newtonian behaviour. This can be expressed mathematically as in Eqn (3.15).

$$\tau = \tau_0 + \eta_B D \tag{3.15}$$

To take into account the extra pseudoplasticity in materials with a flow boundary, Eqn (3.16) according to Casson is often used.

$$\sqrt{\tau} = \sqrt{\tau_0} + \eta_C \sqrt{D} \tag{3.16}$$

For the sake of simplicity, for pseudoplastic or dilatant fluids the apparent viscosity is defined as in Eqn (3.17).

$$\eta_a = \frac{\tau}{D} \tag{3.17}$$

For Newtonian fluids the apparent viscosity is equal to the velocity constant. In non-Newtonian fluids which can be described by Eqn (3.14) it is a function of the shear

Table 3.1   Viscous properties of fluids.

Newtonian fluids	η = const. for p, T const.

Non-Newtonian fluids	• purely viscous: $\eta = f(D)$
	• rheopex: $\eta = f(D, t)$;
	   increases with time increase in D
	• thixotropic: $\eta = f(D, t)$;
	   decreases with increase in D
	• viscoelastic: energy storage;
	   possibility of a reverse flow

stress and Eqn (3.18) is valid.

$$\eta_a = \frac{\tau}{D} = \frac{K_c D^m}{D} = K_c D^{m-1} \tag{3.18}$$

Description of the Experiment

Principle

The measurement of the viscosity of a fluid is generally carried out with a capillary, ball and cylinder, or a rotational viscosimeter. A common form of the ball and cylinder viscosimeter is the Höppler viscosimeter in which the radius of the ball and the cylinder are only minimally different and the movement of the ball is a sliding down the tilted cylinder (Fig. 3.3). This instrument is calibrated with fluids of known viscosity. One obtains a constant value for the instrument, with which the dynamic viscosity may be calculated from the time taken for the ball to travel between two marked positions on the cylinder and the difference in the densities of the ball and the fluid. For the determination of the flow properties of non-Newtonian fluids a ball viscosimeter is not the ideal instrument, because the adjustment of different shear stresses is complicated. Rotational viscosimeters are more suitable and are therefore used more frequently for this purpose. In this instrument the sample fluid is located in the annular clearance between two co-axial cylinders (see Fig. 3.3); one of the two cylinders is set to rotate and the energy transferred is determined by torque measurement on the inner cylinder. If the inner cylinder rotates the instrument is known as a Searle type, and as a Couette viscosimeter if the outer cylinder rotates. By using various speeds of rotation a wide range of shear stresses can be obtained in the annular clearance, thus making it the preferred system for determining the flow properties of non-Newtonian fluids.

Prerequisites for the interpretation of the experiment and calculation of the flow parameters are:

• laminar flow layers
• stationary flow
• adhesion to the wall (no sliding between the sample and the cylinder)
• homogeneous sample (no separation of the sample, e.g. due to sedimentation).

When using a rotational viscosimeter of the Searle type the Taylor number (slit width d, circular velocity u_i and radius R_i of the inner cylinder) must be taken into consideration to ensure that laminar flow layers exist (Eqn (3.19)).

$$Ta = \frac{u_i d\rho}{\eta} \sqrt{\frac{d}{R_i}} \qquad (3.19)$$

In this measurement system the region of laminar flow is left before turbulence sets in. This is because the acceleration of the fluid on the inner cylinder generates a force directed outwards. This causes laminar ring disturbances (Taylor disturbance) which, because of extra convectional transport, permit no further measurement of viscosity. Plotting the resistance number C_m against the Taylor number shows the region of laminar layers, laminar Taylor disturbances and turbulence in the annular clearance (Fig. 3.4). The geometry of the annular clearance in the rotational viscosimeter allows the curve of the velocity profile and the displacement of the laminar flow to be determined analytically and predicted. Such a system of measurement is termed an *absolute system*. Systems with a more complicated geometry (e.g. helical or turbines), which do not permit such an exact calculation, are termed *relative systems*. The correct choice of measuring system is dictated by certain properties of the sample, e.g.:

- approximate value of the viscosity
- homogeneity of the sample
- maximum possible shear force
- application of the procedure.

Materials

VISCOSIMETER

A rotational viscosimeter with a thermostatically controlled sample vessel is necessary. It should be possible to set the speed of rotation manually, and the instrument should be fitted with a gauge to measure torque.

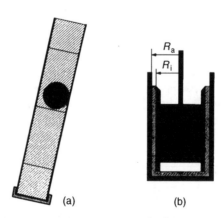

Figure 3.3 Diagrammatic representation of (a) a ball and cylinder viscosimeter and (b) a rotational viscosimeter.

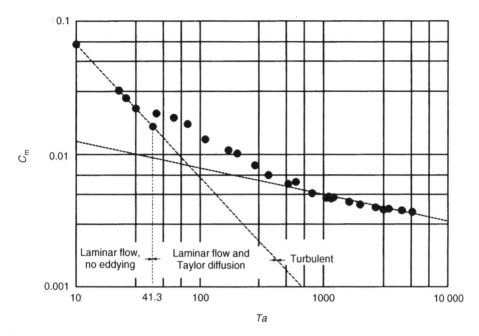

Figure 3.4 Resistance number plotted against Taylor number.

Furthermore some constants must be known to permit the calculation of the shear rate at the annular clearance. A large range of measurements can be obtained by using different torque springs or exchanging the rotational unit and thus altering the annular clearance. A ball and cylinder viscosimeter can be used to obtain comparative measurements but is not essential for the experiment.

FLUIDS

Suitable fluids are:

- glycerol (> 95% pure), which is a high-viscosity Newtonian fluid
- polysaccharide solutions (e.g. xanthane or schizophyllan at about 5 g/l), which are pseudoplastic fluids
- samples of a culture of a mycelium producer (e.g. *Penicillium chrysogenum*), which has pseudoplastic properties.

Procedure

To test the accuracy of the instrument the viscosity of a Newtonian fluid (e.g. glycerol) is measured in the viscosimeter(s) being used for the experiment.

The flow properties of different non-Newtonian fluids are determined in the rotational viscosimeter by choosing a rotational speed, n, and determining the torque required to reach a stationary state. From these two values the change in shear rate, D, the displacement, τ and the viscosity, η, can be calculated.

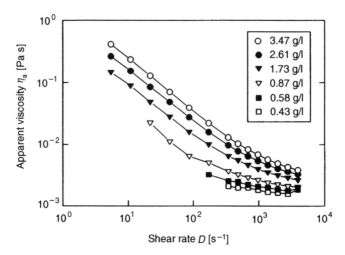

Figure 3.5 Apparent viscosity as a function of shear rate at different concentrations of schizophyllan.

Interpretation, Conclusions and Commentary

The viscosity of a fluid which obeys the displacement equation according to Ostwald and de Waele can be plotted against the shear rate D in a double log plot, and the parameters K_c and m can be determined.

Fig. 3.5 illustrates an example of such a plot for a pseudoplastic polysaccharide. It is important that the experiment is performed at high speeds of rotation (shear rates). Deviation from this equation may occur for two reasons: either the critical value, 41.3, for the Taylor number has been exceeded and the measurements are no longer valid or the viscosity of the solution approaches the limit of the viscosity of the solvent. An aqueous solution was used to produce Fig. 3.5, and the viscosity cannot be less than 10^{-3} Pa s.

Further reading

KULICKE, W.-M. (1986) Fließverhalten von Stoffen und Stoffgemischen (Basel: Hüthig und Wepf Verlag).

SCHLICHTING, H. (1965) Grenzschicht-Theorie, 5th edn (Karlsruhe: Verlag G. Braun).

3.2 REACTION KINETICS

3.2.1 *Determination of Microbial Reaction Kinetics for an Inhibitory Substrate in a Fed-batch System*

P. Götz

Principles and Aims

Knowledge of the dependence of microbial growth rate on the concentration of an inhibitory substrate is particularly important for the running of continuous cultures.

Such a culture can only be maintained at limiting concentrations of the substrate. At concentrations of substrate inhibition there is no stationary stage in a one-step continuous culture at which kinetic data can be obtained. The running of a two-phase continuous culture with the addition of substrate during the second phase is complicated and time-consuming but can be avoided by interpretation of batch experiments. At the start of the batch culture false data about the growth rate at high substrate concentrations can be obtained, because it is difficult to differentiate between the lag-phase and inhibition. Determining the parameters characterising the region of substrate limitation may also pose a problem because this stage of the process is only short-lived and therefore not easy to monitor. The problems outlined above can be avoided by using fed-batch cultures. The aim of this exercise is to determine, on the basis of balanced equations, how the process is to be set up, carry out the relevant experiment and interpret the kinetic data obtained. Setting up a model process to describe the microbial breakdown of an inhibitory substrate requires, among other things, a knowledge of the parameters in the formula for the specific growth rate μ. For the situation in which there is inhibition due to an excess of substrate, Eqn (3.20) (Haldane kinetics) where K_s describes substrate limitation and K_i substrate inhibition is used.

$$\mu = \mu_{max} \frac{S}{K_s + S + \dfrac{S^2}{K_i}} \tag{3.20}$$

To determine the relationship between the specific growth rate μ and the substrate concentration S the growth of the microorganism has to be tested at various substrate concentrations. A simple and time-saving method for doing this is fed-batch cultivation. The process can be described mathematically from the reactor equilibrium, and for biomass X, substrate S and volume V in a fed-batch process with an input rate of F Eqns (3.21) to (3.23) hold.

$$\frac{dX}{dt} = \mu X - \frac{F}{V} X \tag{3.21}$$

$$\frac{dS}{dt} = (S_0 - S) \frac{F}{V} - \frac{1}{Y_{X/S}} \mu X \tag{3.22}$$

$$\frac{dV}{dt} = F \tag{3.23}$$

Using these equations an optimal experiment for determining kinetic parameters can be designed. The aim is to calculate an input programme $F(t)$ which, as a result of interpretation of the quantities X, S and V, gives a reliable estimate of the parameters.

Description of the Experiment

Principle

The kinetics of the breakdown of an inhibitory substrate, e.g. phenol or phthallic acid are to be determined in a fed-batch experiment. To obtain the best possible description of the kinetics, substrate concentrations from the limiting range $(S \to 0)$ and from the inhibiting range $(S \gg 0)$ should be tested. Regular measurement of the variables of the system X, S and V allows an estimate of the growth rate as a function of the substrate concentration, and this is then used to determine the kinetic parameters.

The design of an experiment to determine kinetic parameters can be made with the aid of a sensitivity analysis. The most important information from such an analysis is that a parameter describing a process or state can only be determined in the experiment in which this process or state can be observed. This statement may seem trivial; however, since the mathematical interpretation of unsuitable experiments apparently produces useful data it is essential that experiments are examined for the information they provide.

Organism, Media and Materials

The yeast *Trichosporon beigelii* can be used to determine growth kinetics during the breakdown of phenol or phthallic acid. The solutions A, B and C, described in Table 3.2, used for the cultivation of the yeast are sterilised separately at 121 °C for 20 min to prevent any of the constituents precipitating out. After cooling, the three solutions are combined aseptically. Sterile 1 l Erlenmeyer flasks are filled with 200 ml of the sterilised medium, inoculated with cells taken from an agar slant of the yeast and incubated with shaking (120 rpm) at 30 °C for 24 h.

For the inoculation of the fermenter, one Erlenmeyer flask (2 l starting volume) in the fermenter is required.

Determination of the Biomass Concentration

The dry weight of the biomass is determined by filtration of a sample from the fermenter, drying the filter and weighing it. The concentration of the biomass is determined from the increase in weight of the filter and the volume of the sample taken. It is not possible to determine the biomass photometrically, because the morphology of *Trichosporon beigelii* is dependent on the growth conditions. In

Table 3.2 Composition of the media. The concentrations given are the final concentrations obtained after combining the three separate solutions.

		Shake flask	Bioreactor
Solution A	$MgSO_4 \times 7\ H_2O$	0.5 g/l	0.25 g/l
	$CaCl_2 \times 2\ H_2O$	0.3 g/l	—
Solution B (pH adjusted to 7.0 for flasks)	KH_2PO_4	1.6 g/l	0.8 g/l
	$(NH_4)_2SO_4$	3.0 g/l	1.5 g/l
	Trace elements (Table 3.3)	10 ml/l	10 ml/l
Solution C	Phthalic acid or phenol	0.2 g/l	varying

different experiments single-cell growth as well as mycelial growth can be observed which can, under certain conditions, cause the formation of pellets.

Determination of the Substrate Concentration

The concentration of phthallic acid or phenol in the culture medium is determined by HPLC. The apparatus is checked and calibrated using the usual methods of analysing standard solutions.

1 ml/l of the vitamin solution (filter sterilised) described in Table 3.4 is added to the mineral medium described in Table 3.2. Cultivation in the shaken flasks and the fermenter is carried out at 30 °C and pH 5.5.

Procedure

The experiment is carried out in three stages:

1 preculture in shaken flasks
2 cultivation of the biomass in a batch fermenter (O/N)
3 starting the feeding programme for the fed-batch process.

During the cultivation in the shaken flasks several measurements should be made to allow estimations of the breakdown and growth rates and the yield coefficient. A fermenter is prepared for a batch culture using only 70% of the available volume to allow for the additional volume during the fed-batch period. The amount of substrate

Table 3.3 Composition of the trace element solution.

EDTA	5.2 g/l
$FeSO_4 \times 7\ H_2O$	1.5 g/l
$ZnSO_4 \times 7\ H_2O$	0.07 g/l
$MnSO_4 \times H_2O$	0.1 g/l
H_3BO_3	0.06 g/l
$CoSO_4 \times 7\ H_2O$	0.2 g/l
$CuSO_4 \times 5\ H_2O$	0.03 g/l
$NiCl_2 \times 6\ H_2O$	0.03 g/l
$Na_2MoO_4 \times 2\ H_2O$	0.04 g/l

Table 3.4 Composition of the vitamin solution.

biotin	2 mg/l
pantothenic acid, Ca salt	400 mg/l
inositol	2 mg/l
p-aminobenzoic acid	200 mg/l
pyridoxine hydrochloride	400 mg/l
thiamine hydrochloride	400 mg/l
riboflavine	200 mg/l

and the size of the inoculum can be determined from the breakdown rates obtained in the preculture and should be chosen so that on the following morning the substrate has been used up ($S = 0$). The appropriate mathematical relationship can be derived from the equilibrium of the batch fermenter (for purposes of simplicity, μ is assumed to be constant):

$$S_{t=0} = X_{t=0} \frac{(e^{\mu t} - 1)}{Y_{X/S}} \tag{3.24}$$

Using the estimated values for μ from the preculture, and after determining the time t required for the substrate to be used up, the amount of substrate to be added $S_{t=0}$ can be calculated.

During the night the batch culture can be monitored by measuring pO_2. A decrease in pO_2 means an increase in breakdown, and if this is followed by a rapid increase in pO_2 it is indicative of substrate limitation. The feeding programme to be started in the reactor next morning should give rise to an increase in the substrate concentration so that the region of inhibition is reached and then a reduction to $S = 0$. The time course of the substrate concentration curve is:

$$\frac{dS}{dt} > 0 \qquad \frac{dS}{dt} = 0 \qquad \frac{dS}{dt} < 0 \qquad S = 0$$

Reaction time t

The concentration S_0 in the input and the initial flow rate can be calculated from Eqn (3.22) for a given slope of the substrate concentration curve and the values already used for the other parameters of the process. Samples should be taken every 20 min to determine the concentration of biomass and substrate; the volume of the reaction is calculated using the input rate and the sample volume. Figs 3.6 and 3.7 illustrate examples of such a reaction process. To simplify the interpretation it is important that samples are taken regularly and any alteration in the input rate is made only when a sample is taken. The accuracy with which the course of the substrate concentration curve can be predicted improves during the fed-batch experiment because the estimated parameters are being checked and updated constantly. Towards the end of the experiment, as much information as possible should be obtained about the growth in the region of substrate limitation. The concentration must be reduced slowly towards zero (see $t = 5$ h in Fig. 3.7).

Calculation, Conclusions and Commentary

Calculation of the specific growth rates can be determined from a plot of the slopes of the biomass concentration curve (Eqn (3.25)).

$$\mu = \left(\frac{1}{X} \frac{dX}{dt} \right) + \frac{F}{V} \tag{3.25}$$

A growth kinetic $\mu = f(S)$ is obtained by plotting the substrate concentration against the corresponding specific growth rate. The kinetic parameters are obtained by measuring curves or linearising the values. For the determination of μ_{max} and K_s a Hanes plot (S/μ against S) and for the determination of μ_{max} and K_i a Dixon plot ($1/\mu$ against S) are recommended.

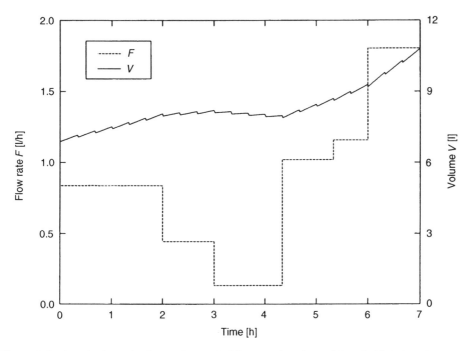

Figure 3.6 Input rate and volumes (corrected for sample volumes) in a fed-batch system.

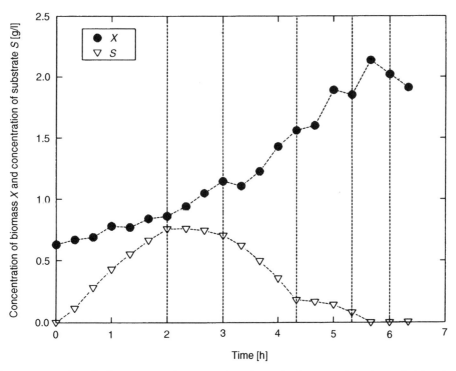

Figure 3.7 Values for biomass (*Trichosporon beigelii*) and substrate (phenol) in a fed-batch system.

Experience shows that a graphical determination of the slope required for Eqn (3.25) can be very inaccurate, therefore this calculation should be done using a numerical method of estimation, e.g. a simplex method coupled to a model simulation of Eqns (3.21) to (3.23).

The influence of dynamic effects (RNA synthesis, enzyme synthesis) on growth in fed-batch cultures can be checked by comparing the values for increasing or decreasing substrate concentrations.

Bibliography

Götz, P. (1992) Einsatz strukturierter Modelle zur Beschreibung dynamischer Zustände mikrobieller Populationen unter Berücksichtigung der biologischen Trägheit. Doctoral thesis, TU Berlin.

3.2.2 Determination of the Parameters of Oxygen Transport in Water

K.-H. Wolf

Aim of the Experiment

The aim of the experiment is to investigate the oxygen uptake capacity of water (as a model medium for a real fermentation medium) as a function of the working parameters, rotation speed of the stirrer and the rate of aeration. This method is to be recommended if one wants to investigate independently of the reaction process the essential hydrodynamic effects on material transport.

The $k_L a$ value and the saturation concentration c_0^* will be determined dynamically under model conditions (water, no reaction) for variable working conditions – speed of stirrer, n, and gas flow rate $\dot{V}_G$. The saturation curves obtained with the pO_2 sampling system are stored on a computer and used to determine the $k_L a$ and c_0^* values by means of non-linear regression.

Theory

Transport between Gas and Liquid Phases

In submerse cultures of aerobic microorganisms (the cells are suspended in the culture medium) there must be a constant adequate concentration of dissolved oxygen in the fermentation medium so that the amount of available oxygen is not a limiting factor for cell growth and the desired metabolic status can be maintained. This is achieved by defined working conditions ($\dot{V}_G$, n). The concentration of dissolved oxygen attained is a function of the oxygen uptake rate (OUR) of the cells and the rate of oxygen transfer (OTR) from the air bubble to the cell. The transfer occurs in a series of steps:

1 transport from within the air bubble through the gas boundary layer to the interface

2 transport across the surface of the interface

3   transport from the surface of the interface through the boundary layer of the liquid into the liquid

4   transport through the liquid

5   transport through the laminar liquid boundary layer at the cell surface

6   transport within the cell.

The amount of material being transported from the gaseous phase can for the simplest and most frequent example be expressed as a first-order reaction based on the serial switching of partial resistances k_i. For each of these steps the general equation of motive force can be applied:

$$\dot{n} = k_i\, a(c_i - c_0) \tag{3.26}$$

Henry's law links the equilibrium constant in the liquid, c_0^*, to the molar fraction in the air space at the gas/liquid boundary:

$$c_G^* = \frac{H}{p}\, c_0^* \tag{3.27}$$

where p is the absolute pressure of the gaseous phase, G gaseous form, H Henry constant, 0 liquid, air at 37 °C: $H = 95\,040$ Pa m^3/mol.

According to the double-film theory the matter boundary gaseous/liquid can be described such that on both faces of the phase boundary surfaces a thin film is formed through which the oxygen is transported by means of molecular diffusion. Both coefficients of conversion k_G (on the gaseous face) and k_0 (on the liquid face) can be expressed as quotients of the coefficients of diffusion (D_G, D_L) at the gaseous or liquid faces and the respective thickness of the film δ_G and δ_L:

$$k_i = \frac{D_i}{\delta_i} \tag{3.28}$$

The rate-limiting step for the entire transport process is the step with the highest resistance. Expressing the material transport in terms of the general equation for motive force, Eqn (3.26), only the conversion coefficient at the liquid face, k_0, is important, since the coefficient of diffusion at the gas face, D_G, is 10^4 times greater than that at the liquid face and the resistance to transport at the gas face can be neglected. For substances with a low viscosity the second transport step at the gas/liquid interface is rate limiting, i.e. relevant for the model and the procedure. Equation (3.29) is valid:

$$\dot{n} = k_L\, A(c_0^* - c_0) \tag{3.29}$$

where $\dot{n}$ is oxygen flow, A surface area at the interface, c_0^* saturation concentration of O_2, and c_0 the actual O_2 concentration. If Eqn (3.29) is expressed relative to the liquid volume V_L Eqn (3.30) is obtained:

$$\text{OTR} = \frac{\dot{n}}{V_L} = k_L\, \frac{A}{V_L}\, (c_0^* - c_0) = k_L\, a(c_0^* - c_0) \tag{3.30}$$

where a is the specific area of the interface.

The specific area of exchange (= area of interface) is related to the Sauter diameter d_B and the volume of the gaseous phase ε_G (gas hold-up) as shown in Eqns

(3.31) to (3.33).

$$a = \frac{A}{V} \approx \frac{6(1 - \varepsilon_G)}{d_B} \tag{3.31}$$

where d_B is the average diameter of the air bubbles, and a is the average space (area) occupied during the process.

$$\varepsilon_G = \frac{V_G}{V_R} = \frac{V_G}{V_L + V_G} = \frac{V_R - V_L}{V_R} = 1 - \frac{V_L}{V_R} = \frac{H_G}{H} \tag{3.32}$$

where H_G is the height of aeration in a fermenter with a level base and H is the height of the liquid in the fermenter.

$$d_B = \frac{\sum_i n_i x_i^3}{\sum_i n_i x_i^2} \tag{3.33}$$

where n_i is the number of bubbles i, x_i the diameter of bubble i.

Volumetric Oxygen Transfer Coefficient $k_L a$

Since k_L and a are only calculated individually in exceptional cases and it is very difficult to measure them separately, they are combined to give the volumetric oxygen transfer coefficient $k_L a$. This complex coefficient is relatively easy to obtain.
The $k_L a$ value is affected by the following factors:

- running conditions (speed of stirrer, gas flow)
- geometry of the reactor (stirring system, capacity, method of aeration)
- material (diffusion constant, surface tension, specific gravity, viscosity, presence of dissolved substances, e.g. salts, glucose, etc.).

There are several correlations in the literature which can be used for calculating the $k_L a$ value; most of them are valid for only a narrow range of values. The most frequently used are the formulae from similarity theory:

- Zlokarnik (1978) – volume-dependent method – Eqn (3.34)

$$k_L a = c_0 \left(\frac{P}{V_L} \right)^{c_1} \left(\frac{\dot{V}_G}{V_L} \right)^{c_2} \tag{3.34}$$

and

- Henzler (1982) – area-dependent method – Eqn (3.35)

$$k_L a = c_0' \left(\frac{P}{V_L} \right)^{c_3} W_G^{c_4} \tag{3.35}$$

where P is the energy input, $\dot{V}_G$ is the gas throughput, V_L is the working volume of the fermenter, W_G is the speed of the stirrer in the empty fermenter, $W_G = \dot{V}_G / A = \dot{V}_G / (\pi d_1^2 / 4)$.

Description of the Experiment

Principle

There are a number of methods, differing in their theoretical basis, technical requirements and practical input, for the experimental determination of the volume-dependent coefficient of transport $k_L a$. One of these methods, the dynamic, requires little technical equipment and can be performed in both 'real' fermentation medium or under model conditions. The principle of the dynamic method for determining $k_L a$ is based on the assumption of an ideally aerated stirred vessel, in which the actual oxygen concentration is altered by giving oxygen signals to the medium. At time t the aeration is stopped (aeration shut off), and after the oxygen concentration has fallen (due to uptake by the microorganisms) it is switched on again. During this time the amount of dissolved oxygen is measured with a pO_2 electrode.

For the ideal stirred vessel the oxygen equilibrium is expressed as shown in Eqn (3.36).

$$\frac{dc_0}{dt} = \text{OTR} - \text{OUR} = \underset{\text{rate of saturation}}{k_L a(c_0^* - c_0)} - \underset{\text{rate of depletion}}{R_0} \tag{3.36}$$

PHASE I (EQUILIBRIUM PHASE) (SEE FIGS 3.8 AND 3.9)

In phase I $(t < t_{02})$ there is a constant equilibrium concentration c_{0S} achieved by the balance of the rate of oxygen transport (OTR) and the rate at which oxygen is used up (OUR). Eqn (3.37) is valid:

$$\frac{dc_0}{dt} = 0 \tag{3.37}$$

Eqn (3.38) follows from Eqn (3.36).

$$k_L a = \frac{R_0}{c_0^* - c_{0S}} \tag{3.38}$$

Since the rate at which the oxygen is used up (R_0) and the saturation concentration (c_0^*) are unknown the value of $k_L a$ cannot be determined.

A transition region (T_1) follows Phase I (stationary phase) in which, after switching off the air flow to the fermenter, the remaining air bubbles in the fermenter medium represent a reservoir of oxygen. When the bubbles have risen to the top of the fermenter Eqn (3.39) is valid for OTR = 0

$$\frac{dc_L}{dt} = -\text{OUR} = -R_0 \tag{3.39}$$

PHASE Ii (CONSUMPTION PHASE) (SEE FIGS 3.8 AND 3.9)

Integration of Eqn. (3.39) gives a value for n within useful limits:

$$c_0 = c_0^* - R_0(t - t_{02}') \tag{3.40}$$

After linear regression in the range $t_{02}' \leqslant t \leqslant t_{03}$ the data obtained in Phase II give values for c_0^* and R_0.

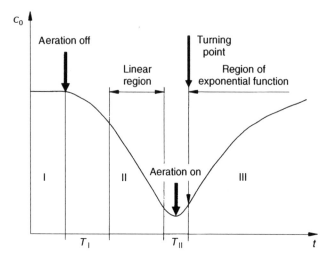

Figure 3.8 Oxygen concentration versus time for the dynamic method.

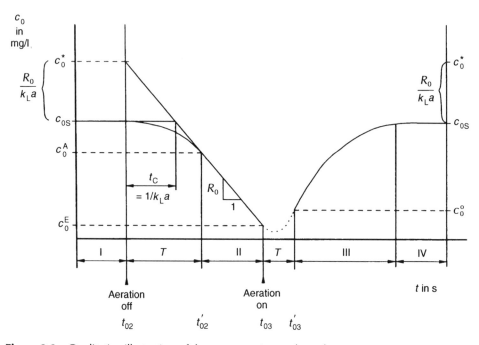

Figure 3.9 Qualitative illustration of the consumption and uptake curves.

If the oxygen concentration falls to the critical value c_{crit} the aeration must be switched on again. There now follows a second transition region T_{II} in which the oxygen concentration of the solution falls initially to the minimum value c_{min} because the air introduced cannot reach all parts of the fermenter immediately (Fig. 3.8). There is therefore a delay in the start of the saturation phase (Phase III). Eqn (3.36) is valid for *Phase III (saturation phase)*. The data curve obtained taking into consideration the valid equlibria for the different phases is shown in Fig. 3.9.

In the model system used for the experiment to determine $k_L a$ in water, i.e. *in the case where no reaction is taking place*, the following simplifications are valid:

- $R_0 = 0$ because there is no O_2 sink in the system
- only data for uptake (see Fig. 3.11) will be considered.

The concentration c_0^E can be achieved by flushing the system with nitrogen. At time t_{03} the degassing with nitrogen is stopped and from time t'_{03} there is constant aeration. The next phase (Phase III) is one of uptake. The curve obtained for $c_0 = f(t)$ is followed until there are sufficient constant values at the upper end of the curve. For this event Eqn (3.41) is valid:

$$\frac{dc_0}{dt} = \text{OTR} = k_L a(c_0^* - c_0) \tag{3.41}$$

Solving the differential equation (3.41) for the starting conditions $c_0(t = t_0) = c_0^o$ results in Eqn (3.42).

$$c_0(t) = c_0^* - (c_0^* - c_0^o)\exp[-k_L a(t - t'_{03})] \tag{3.42}$$

In principle there are two ways of determining the $k_L a$ value and c_0^* from the curve $c_0(t)$ obtained:

1 conversion of Eqn (3.41) into a differential equation and performing a linear regression (Wolf, 1991)

2 non-linear regression with Eqn. (3.42) and determining the parameters $k_L a$, c_0^* and c_0^o.

To determine the $k_L a$ value by the dynamic method, short bursts of measurement with the electrode are necessary so that measurement time with the electrode has no influence on the value of $k_L a$.

These conditions are fulfilled by the electrodes named below. Furthermore a sufficiently high rate of current must be applied otherwise air bubbles collect on the electrode and cause false readings to be made. This means that for $k_L a$ values > 0.1 s^{-1} (characteristic time constant $t_c = 1/k_L a > 10$ s; see Fig. 3.9), the time course of the pO$_2$ electrode must be taken into consideration. The dynamic method is suitable as long as $k_L a \leqslant 0.3$ s^{-1}.

Media and Materials

The experimental set-up is shown in Fig. 3.10. The pO$_2$ electrode used is a TriOxmatic 103. Tap water is used as the medium. Aeration is provided by an aeration system below the stirrer. Interchangeable valves allow compressed air or nitrogen to be used for aeration or degassing the system. The fermenter is thermostatically controlled.

Procedure

PREPARATION

1 Calibrate the pO$_2$ electrode according to the manufacturer's instructions.

2 Set the temperature of the fermenter to 20 °C.

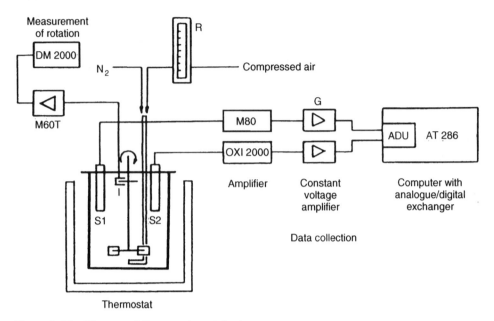

Figure 3.10 Diagram of the experimental set-up.

3 Position the pO_2 electrode so that the membrane is at the height of the stirrer, set the electrode to measure concentration (mg/ml) and allow the electrode to reach the temperature of the fermenter.

4 Switch on and set data collector to measure every 5 s.

EQUIPMENT USED

Data collection:

S1	pO_2 electrode, reference electrode
S2	pO_2 electrode
M80	amplifier (output voltage $0-1$ V)
OXI 2000	oxygen meter (output voltage $0-0.5$ V)
G	direct current amplifier; amplification: 20 dB for M80, 30 dB for OXI 2000
ADU	analogue to digital exchanger, input voltage $1-10$ V
AT 286	computer with AT 286 processor.

Measurement of stirrer speed:

I	regulator, perforated disc mounted on the stirrer (20 perforations) which interrupts the light beam emitted from the light diode and received by the photo diode
M60T	transistorised amplifier, 600 Ω connection, 20 dB amplification
DM 2000	frequency meter.

Flow rate measurement:

R   flow meter (calibrated for air at 20 °C and 760 torr normal air
              pressure)

              range of measurement 1 : 10–90 l/h (±3 l/h)

              range of measurement 2 : 50–259 l/h (±5 l/h).

- Enter a calibration value into the data collection system for defined conditions:

  $\dot{V}_G = 0$ l/h

  $n = 270$ min^{-1} (= 90 impulses/s)

  $T = 20$ °C

- Laboratory fermenter, $V_L \approx 2$–5 l made of glass and with stirrer speed control. If a computerised small-scale fermenter is available, so much the better, because all the extra measuring devices described above are part of the fermenter and all data collected are automatically stored and processed.

DETERMINATION OF THE OXYGEN UPTAKE CURVES

It is recommended that the data for producing the oxygen uptake curves are obtained under the following conditions:

- $n = 390$ min^{-1} kept constant

  $\dot{V}_G = 60, 80, 100, 130, 160$ l/h (all values should be measured in the range 50–350 l/h)

- $\dot{V}_G = 80$ l/h constant

  $n = 290, 340, 440, 490$ min^{-1}

1 Connect compressed air to the aeration valve and set the flow meter to $\dot{V}_G$.

2 Connect up the tubing to allow the fermenter to be flushed with nitrogen; interrupt flow of compressed air.

3 Flush fermenter with nitrogen; the amount of oxygen registered on OXI 2000 should be between 0.1 and 0.09 mg/l.

4 Ensure that the data are registered in the data collector.

5 After flushing with nitrogen: switch off N$_2$ supply, stop the stirrer (any bubbles of N$_2$ on the stirrer which would interfere with the aeration rise to the top of the fermenter), select desired rotation speed, switch on the data collector and open the aeration valve. Note the time at which aeration started.

6 Immediately after the onset of aeration adjust the speed of stirring. (In the aerated fermenter the energy transfer falls and as a consequence the speed of rotation increases.) Further control and adjustment will be necessary throughout the experiment.

7 The experiment can be stopped as soon as constant measurements have been obtained over a period of 30 s, i.e. there has been no change in the first decimal place of the values obtained.

It is practical to do all the experiments for a particular rate of aeration before the next flow rate is set.

Calculation, Interpretation and Conclusions

The values for $k_L a$ and c_0^* are calculated using appropriate software. Software is available which allows chosen models, e.g. Eqn (3.42) to be included in the analysis.

Fig. 3.11 illustrates the curves obtained between 0 and t. In the initial stages the curves have a distinct sigmoid shape with a turning point, t_w, caused by a variety of reasons. Calculation of the data according to Eqn (3.42) with the initial values $c_0(t_0')$ describes only one e-function. Therefore a regression or parameter is determined which is in principle wrong. The error can be dealt with in either of two ways:

1   The starting point is placed after c_0' at the turning point t_w (this involves an exact estimation of the terms $c_0' \neq 0$, $t_0 \neq 0$ from the monitor). This method has the disadvantage that data are lost from the initial readings obtained from the electrode.

2   A reply function is used which takes the time behaviour of the electrode into consideration.

It has been shown that both methods give almost identical results; however, since the latter method requires complicated calculations, the former will be used here.

Eqn (3.42) is used to calculate the concentration curves. The individual values are as follows:

Variable: $c(t)$, t

Parameter: $k_L a$, c_0^*, c_0'

Constants: $t_0 = t_{03}' \neq 0$

The relevant initial estimates and limits are entered as shown in Table 3.5. All concentration curves are to be calculated according to this method. The following values should be plotted against each other on double log paper:

$k_L a = f(n)$ parameter: $\dot{V}_G$

$k_L a = f(\dot{V}_G)$ parameter: n

Comment on the dependencies taking Eqns (3.34) and (3.35) into consideration.

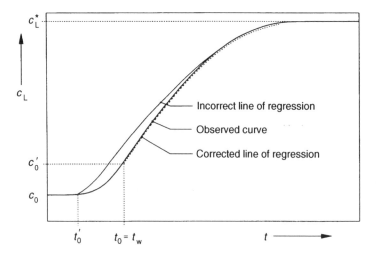

Figure 3.11   Positions of the regression curves.

Table 3.5 Initial estimates and limits used in calculation of concentration curves.

Initial estimates		Limits	
		Minimum	Maximum
$k_{L}a$ [s^{-1}]	0.02	10^{-3}	10^{-1}
c_0^{*} [mg l^{-1}]	7	5	10
c_0' [mg l^{-1}]	1	0.1	3

Bibliography

GOLDSCHMIDT, B., LINDNER, U., MATHISZIK, B. and TILLER, V. (1992) Modellbank Biotechnologie V 6.1 Softwarepaket zur Erfassung, Verarbeitung und Auswertung von Daten biochemischer, chemischer und physikalischer Versuche. Martin-Luther-Universität, Halle-Wittenberg, Institut für Biotechnologie.

HENZLER, H.-J. (1982) Verfahrenstechnische Auslegungsunterlagen für Rührbehälter als Fermenter. *Chem.-Ing.-Tech.* **54**, 461–76.

SCHLÜTER, V. (1992) Charakterisierung und Maßstabsübertragung von Biorührreaktoren. *Fortschr.-Ber. VDI*, Reihe 3, Nr. 285 (Düsseldorf: VDI-Verlag).

STEJSKAL, U. (1992) Stofftransport ohne Reaktion in Wasser. Diplomarbeit (DA-Nr.: 775), Fakultät für Maschinenwesen, Institut für Bioverfahrenstechnik, TU Dresden.

WOLF, K.-H. (1991) *Berechnungsbeispiele zur Bioverfahrenstechnik* (Hamburg: Behr's Verlag).

ZLOKARNIK, M. (1978) Sorption characteristics for gas–liquid contacting in mixing vessels. *Adv. Biochem. Eng.* **8**, 133–51.

3.2.3 Determination of the Deactivation Kinetics of Soluble Enzymes

H. Voß

Principle of the Experiment

An important criterion when using enzymes in technical processes for material conversion, or in analytical procedures, is the stability of the enzyme during the process. The stability of the enzyme dictates the length of time for which the enzyme is active under the reaction conditions. Depending on the type of enzyme, its purity, enzyme pretreatment and reaction conditions, the enzyme activity can change detectably over periods of time ranging from minutes to years. The main factors affecting enzyme stability are: temperature, pH, concentrations of substrate and product, chemical effects (inhibitors, heavy metals, impurities), breakdown caused by microbes.

In the case of immobilised enzymes the reagents used for immobilisation, release from the carrier, mechanical wear and possible diffusion effects all influence the stability of the enzyme(s). Investigation of enzyme deactivation is not only useful for determining the productivity Pr (g/l h) for a given amount of enzyme and the half life $t_{1/2}$, i.e. the time it takes for the initial activity of the enzyme to drop by

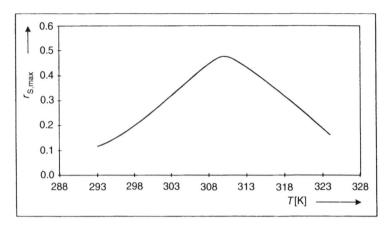

Figure 3.12   Temperature dependency of the catalytic activity of an enzyme: $E_A = 80$ kJ/mol, $E_D = 200$ kJ/mol, $B = 2.8 \times 10^{33}$, $A\,c_{E0} = 2.28 \times 10^{13}$.

one-half, but also for explaining the causes of deactivation and possible ways of influencing it.

Temperature Dependency of the Catalytic Activity and Stability of an Enzyme

If the enzyme reaction follows a Michaelis–Menten relation

$$r = r_{max} \frac{C_S}{K_S + C_S} \tag{3.43}$$

then the maximum velocity of the reaction increases initially with temperature according to the Arrhenius equation

$$r_{max} = A c_E \exp\left(-\frac{E_A}{RT}\right) \tag{3.44}$$

where A is the pre-exponential factor, c_E the enzyme concentration, E_A is the activation energy and R is the universal gas constant. At higher temperatures the conformation of the enzyme is altered because of stronger rotational forces and movement of the molecules, and this in turn has an adverse effect on the ability of the enzyme to catalyse the reaction. As illustrated in Fig. 3.12, this results in a fall in the activity of the enzyme. The optimal reaction temperature for technical enzymes lies between 20 °C and 60 °C. If this temperature is exceeded only minimally then the conformational changes are reversible and the deactivation can be described thermodynamically:

$$E_{native} \rightleftharpoons E_{denatured} \tag{3.45}$$

If the equilibrium sets in quickly, the temperature dependence of the denaturing equilibrium constant K_D can be termed as shown in Eqns (3.46) to (3.48)

$$K_D = \frac{c_D}{c_N} \tag{3.46}$$

$$c_N = \frac{c_{E0}}{1 + K_D} \tag{3.47}$$

$$K_D = \exp\left(\frac{\Delta S_D}{R}\right)\exp\left(\frac{-E_D}{RT}\right) \tag{3.48}$$

where c_D is concentration of denatured enzyme, c_N is concentration of native enzyme, ΔS_D is denaturing entropy (kJ/mol K) and E_D is deactivation energy under standard conditions.

In summary Eqns (3.44) to (3.48) give the connection between enzyme activity and temperature illustrated in Fig. 3.12.

$$r_{max} = A\,\frac{c_{E0}\,\exp(-E_A/RT)}{1 + B\,\exp(-E_D/RT)} \tag{3.49}$$

where B is the entropy term of Eqn (3.48).

Kinetics of Deactivation

Irreversible deactivation processes are described by kinetic expressions. Enzyme deactivation is a reaction of the first order:

$$E_N \rightarrow E_D \tag{3.50}$$

(E_N is native enzyme, E_D is deactivated enzyme) with the rate equation

$$\frac{da}{dt} = -k_d a \tag{3.51}$$

where a is activity and k_d is rate of deactivation. Integration results in Eqn (3.52).

$$a = a_0\,e^{-k_d t} \tag{3.52}$$

As shown in Fig. 3.13, deactivation is extremely temperature dependent according to Eqn (3.53).

$$k_d = k_{d0}\,e^{-E_{A,d}/RT} \tag{3.53}$$

The production in the fermenter is represented by the area under the curve. Over a short reaction period (<10 h) the higher temperature is to be recommended, but for longer periods the lower is preferable. The relationship between half life, $t_{1/2}$, and the deactivation constant is given by Eqn (3.54).

$$t_{1/2} = \ln 2/k_d \tag{3.54}$$

Other kinetics of enzyme deactivation are to be found in Henley and Sadana (1985).

Aims

From the measurements of enzyme activity at various times and different temperatures the order of the deactivation kinetic, the rate constant of deactivation, and the activation energy of the deactivation will be calculated.

Firstly, it has to be determined whether deactivation is a reaction of the first order or if it follows another course. A reaction is of the first order if the plot of $\ln(a/a_0)$

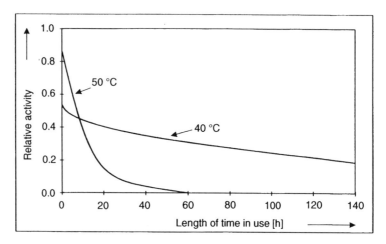

Figure 3.13 Activity of glycoamylase as a function of period of use at two different temperatures.

against time gives a straight line (Fig. 3.14):

$$\ln(a/a_0) = -k_d t \tag{3.55}$$

The slope of the straight line gives k_d.

If $\log_n$ of the various deactivation rate constants $k_d(T)$ is plotted against the reciprocal of the absolute temperature, the slopes of the straight lines obtained give the negative quotient of the deactivation energy (E_A) and the gas constant $(R = 8.314 \text{ J mol}^{-1} \text{ K}^{-1})$ (Fig. 3.15):

$$\tan \alpha = -E_{A,d}/R \tag{3.56}$$

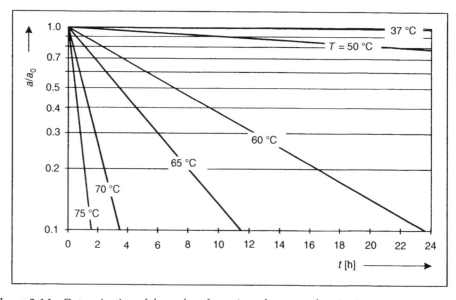

Figure 3.14 Determination of the order of reaction of enzyme deactivation.

Numerically, the relation given by Eqn (3.57) can be stated.

$$k_d(T) = k_{d,0} \, e^{-E_{A,d}/RT} \qquad (3.57)$$

where $k_{d,0}$ is the so-called pre-exponential factor or the action constant which may be expressed as shown in Eqn (3.58).

$$k_{d,0} = k_d(T)/e^{-E_{A,d}/RT} \qquad (3.58)$$

If these terms are stated for two different temperatures with the corresponding k_d values as measured, the deactivation energy can be calculated as follows.

$$k_d(T_1) = k_{d,0} \, e^{-E_{A,d}/RT_1} \qquad (3.59)$$

$$k_d(T_2) = k_{d,0} \, e^{-E_{A,d}/RT_2} \qquad (3.60)$$

These equations can be combined to give Eqn (3.61).

$$k_d(T_1) = k_d(T_2) e^{-E_{A,d}/R(T_1 - T_2)} \qquad (3.61)$$

From Eqn (3.61) it follows that:

$$\ln\left(\frac{k_d(T_1)}{k_d(T_2)}\right) = \frac{-E_{A,d}}{R}\left(\frac{1}{T_1} - \frac{1}{T_2}\right) \qquad (3.62)$$

$$E_{A,d} = \ln\left(\frac{k_d(T_1)}{k_d(T_2)}\right) R \left(\frac{1}{T_1} - \frac{1}{T_2}\right)^{-1} \qquad (3.63)$$

The problems to be solved may be summarised as follows:

1   determination of the activity of the free enzyme at various temperatures and times of incubation

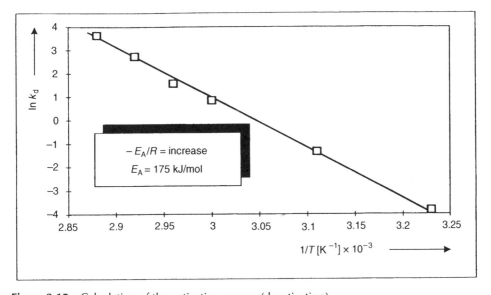

Figure 3.15   Calculation of the activation energy (deactivation).

2   calculation of the order of reaction for the deactivation kinetics at a specific temperature

3   calculation of the rate constant for the deactivation reaction

4   calculation of the energy of deactivation, the action constant and the half life of the enzyme at three different temperatures.

Description of the Experiment

Principle

The temperature-dependent deactivation kinetics of catalase (hydrogen peroxidase oxido-reductase, EC 1.11.1.6) will be examined. Catalase splits hydrogen peroxide into water and oxygen:

$$2H_2O_2 \rightarrow 2H_2O + O_2$$

At substrate concentrations above 0.1 M catalase is substrate inhibited. Catalase activity is dependent on the presence of certain metal ions. The optimum pH lies between 6.8 and 7.5 and the turnover number is 6×10^6 mol substrate/(mol enzyme min). The temperature range of activity is given as 25–45 °C, and above 55 °C there is a significant loss in activity.

Unit activity of catalase is defined as follows. One unit (U) splits 1.0 μmol hydrogen peroxide per minute at pH 7.0 at 25 °C, causing the concentration of hydrogen peroxide to drop from 10.3 mM to 9.2 mM.

The turnover rate for H_2O_2 can be measured optically at 240 nm, or the amount of O_2 released can be measured. To calculate the deactivation constants and the activation energy of the deactivation, catalase activity is measured in a batch reactor at various temperatures, and the rates are determined from the amount of O_2 released over a given period of time.

Materials

Catalase (stabilised, 260 000 U/ml at 25 °C [Boehringer Mannheim]); 750 ml 0.01 M acetate buffer, pH 4.5; 1.2 l 0.05 M phosphate buffer, pH 7.0; substrate solution (30% hydrogen peroxide solution in 0.05 M phosphate buffer, pH 7.0); water baths at 25 °C, 40 °C and 60 °C; measuring cylinders (100 ml); beakers, magnetic stirrers, pipettes (1 and 10 ml); twelve 100 ml Erlenmeyer flasks; two bored bungs; plastic tubing, graduated test tubes; Parafilm™.

Procedure

1   180 μl catalase is added to 90 ml 0.01 M acetate buffer, pH 4.5. 10 ml of this catalase dilution is placed into each of nine 100 ml Erlenmeyer flasks.

2   To determine the activity of catalase, 100 ml H_2O_2 solution is added to each flask. The flask is stoppered with a bored bung with plastic tubing going from the mixture in the flask into an inverted graduated test tube immersed in a beaker of water. After 5 min the volume of water displaced by the oxygen generated by the catalase is noted. The enzyme activity is determined from the amount of oxygen produced as described in step 3.

Table 3.6 Plan for testing catalase activity.

No.	T (°C)	Activity after	
		h	U
1	25	0	
2	25	4	
3	25	6	
4	40	1	
5	40	4	
6	40	6	
7	60	1	
8	60	4	
9	60	6	

3 *Example*: 1 ml O_2 is produced per minute.

   1 mol O_2 = 22 400 ml

   45 μmol O_2 = 1 ml

   i.e. 45 μmol oxygen is produced from 90 μmol H_2O_2. The activity of the catalase is therefore 90 U.

4 The nine samples are incubated at the three temperatures and the times shown in Table 3.6.

Interpretation

The enzyme activities measured at the various temperatures are the initial rates. When the activities are plotted according to Eqn (3.55) the $k_d(T)$ values are obtained. The $E_{A,d}$ values are obtained by plotting the values according to Eqn (3.56).

Reference

HENLEY, J.P. and SADANA, A. (1985) Categorization of enzyme deactivation using a series type mechanism. *Enz. Microb. Technol.* **7**, 50–7.

3.2.4 *Deactivation of Immobilised Enzymes Subject to Diffusion*

H. Voß

Principle and Aim of the Experiment

The activity of enzymes immobilised on porous carriers may be subjected to internal diffusion. In this instance the resistance to diffusion affects the apparent rate of deactivation in that the effective deactivation is reduced as the resistance to diffusion increases. Therefore it is necessary to know the effect of diffusion on the activity/time behaviour of an enzyme when calculating the useful time or the half life of immobilised enzymes.

Another aspect of immobilisation of enzymes is that very often there is a reduction in the initial activity of the immobilised enzyme as compared with the soluble enzyme and this is coupled with a simultaneous increase in the effective stability of the immobilised enzyme (Buchholz, 1979). The relationships mentioned in Subsection 3.2.3 for soluble enzymes are also valid for the purely kinetic control of the deactivation process in immobilised enzymes. Assuming Michaelis–Menten kinetics for the reaction catalysed, the following equations are valid:

$$r = r_{max} \frac{C_S}{K_m + C_S} \qquad (3.64)$$

$$r_{max} = (k_2 a)\, c_E \qquad (3.65)$$

where r_{max} = maximum reaction velocity, $(k_2 a)$ = product of the constants of the rate-limiting step and the specific activity of the immobilised enzyme [1/(gs)] and c_E = concentration of the immobilised enzyme in the reaction medium (g/l).

If diffusion influences the enzyme reaction as shown in Fig. 3.16, then there is an effective, measurable rate of mass change R_S.

$$R_S = \frac{dC_S}{dt} = \eta r \qquad (3.66)$$

In the above equation, η is the *effectivity factor* or the degree of pore utilisation.

$$\eta = \frac{\text{reaction rate taking diffusion into account}}{\text{reaction rate ignoring diffusion}} \qquad (3.67)$$

The connection between material transport and enzyme kinetics can be described as follows:

$$\eta = f(\phi) \qquad (3.68)$$

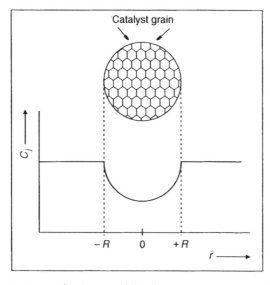

Figure 3.16 Concentration gradients: immobilised enzyme–carrier-catalyser.

where ϕ is the term for diffusion or the Thiele term. The Thiele term is defined for a reaction of quasi-first order $(K_m \gg c_S)$ as:

$$\phi = \frac{R_P}{3} \sqrt{\frac{r_{max}}{K_m D_e}} \tag{3.69}$$

where R_P is the radius of the particle and D_e is the coefficient of effective diffusion. Fig. 3.17 illustrates the dependency of the effectivity factor on the Thiele term – i.e. primarily on the size of the catalyser particle defined as R_P.

To determine the influence of particle size of the enzyme carrier, and therefore diffusion, on the kinetics of deactivation, an *instability factor*, γ is defined (Eqn (3.70)).

$$\gamma = \frac{\text{diffusion-dependent rate of deactivation}}{\text{diffusion-independent rate of deactivation}} \tag{3.70}$$

As in the experiment described in Subsection 3.2.3, a deactivation with first-order kinetics is assumed.

$$\frac{da}{dt} = k_d a \tag{3.71}$$

therefore:

$$k_d = \frac{d(\ln a)}{dt} \tag{3.72}$$

Correspondingly, Eqns (3.64) and (3.65) are valid.

$$\frac{d(\ln r)}{dt} = \frac{d(\ln a)}{dt} = k_d \tag{3.73}$$

Differentiating Eqn (3.66) with respect to time results in Eqn (3.74).

$$\frac{dR_S}{dt} = \eta \frac{dr}{dt} \left[1 + \frac{d(\ln \eta)}{d(\ln r)} \right] \tag{3.74}$$

Eqn (3.69) states that the rate of the reaction r is proportional to ϕ^2, therefore:

$$\frac{dR_S}{dt} = \gamma \eta \frac{dr}{dt} \tag{3.75}$$

where the instability factor

$$\gamma = 1 + \frac{1}{2} \left[\frac{d(\ln \eta)}{d(\ln \phi)} \right] \tag{3.76}$$

for $(0.5 < \gamma \leqslant 1)$.

According to Eqns (3.66) and (3.75) the following is valid:

$$\gamma = \frac{d(\ln R_S)}{d(\ln r)} \tag{3.77}$$

which is the mathematical version of Eqn (3.70).

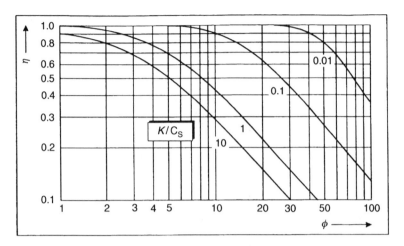

Figure 3.17 Dependence of the effectivity factor on the diffusion term.

Taking Eqn (3.73) into account, the diffusion-reduced deactivation rate is

$$\frac{d(\ln R_S)}{dt} = \frac{d(\ln r)}{dt} = \gamma k_d \tag{3.78}$$

For spherical catalysers and heterogeneous reactions of quasi-first order ($K_m \gg c_S$ in Eqn (3.64)) the following equation is valid for the effectivity factor η (Baerns *et al.*, 1987):

$$\eta = \frac{1}{\phi}\left(\frac{1}{\tanh(3\phi)} - \frac{1}{3\phi}\right) \tag{3.79}$$

and, according to Eqn (3.76),

$$\gamma = \frac{0.5\left(1 - \dfrac{6\phi}{\sinh(6\phi)}\right)}{1 - \dfrac{\tanh(3\phi)}{3\phi}} \tag{3.80}$$

The apparent reaction rate is altered as a result of enzyme deactivation as described in Eqns (3.73) and (3.78). However, when diffusion is playing a role, the alteration is less by the factor γ than when it does not play a role.

The dimensionless unit of time for this difference in deactivation rates is the *apparent stationary time*, θ, of the immobilised enzyme:

$$\theta = \frac{t_R}{t_E} \tag{3.81}$$

where t_R is the reaction time according to Eqn (3.69):

$$\ln\left(\frac{R_S}{R_{S0}}\right) = \ln\frac{\eta}{\eta_0} - k_d t_R = \ln\beta \tag{3.82}$$

and t_E the enzyme deactivation time as described by Eqn (3.72):

$$\ln \frac{a}{a_0} = -K_d t_E = \ln \beta \qquad (3.83)$$

or:

$$t_E = -\frac{\ln \beta}{k_d} \qquad (3.84)$$

and

$$t_R = \frac{\ln \beta - \ln(\eta/\eta_0)}{-k_d} \qquad (3.85)$$

Eqns (3.84) and (3.85) provide the relative apparent stationary time, θ, in the form of measurable activities and effectivity factors η and η_0 (η at time 0):

$$\theta = 1 - \left[\frac{\ln(\eta/\eta_0)}{\ln \beta} \right] \qquad (3.86)$$

Aim of the Experiment

Firstly the constants of the Michaelis–Menten equation, r_{max} and K_m, will be determined for the free enzyme at $T = 40\,°C$. By measuring the starting rates (apparent activities) of the immobilised enzyme as a function of the particle radius and the half life at the same temperature, $40\,°C$, the effectivity factors η_0 and η, the deactivation rate k_d, the instability factor γ and the relative length of the half life θ of the enzyme preparations will be determined.

Description of the Experiment

Using the same experimental set-up as described in Subsection 3.2.3, the Michaelis–Menten kinetics of H_2O_2 breakdown by soluble catalase are determined. Catalase is then immobilised on five different preparations of the cation exchanger carboxymethylcellulose differing in particle size. In batch experiments the starting activity is determined for each particle size at varying times. The values obtained are then used to calculate the effectivity factor η, the instability factor γ, the kinetic deactivation rate k_d and the relative half lives for each of the immobilised preparations.

Equipment and Solutions

Catalase (stabilised, 260 000 U/ml at 25 °C [Boehringer Mannheim]); carboxy-methylcellulose beads with 0.2, 0.31, 0.4, 0.63 and 0.8 mm diameter (CM-cellulose); substrate solution (0.2 ml 30% H_2O_2 solution); 500 ml 0.1 M acetate buffer, pH 4.5; 1200 ml 0.05 M phosphate buffer, pH 7.0; 100 ml measuring cylinder; 1 l beaker; magnetic stirrer; 1 ml and 10 ml pipettes; ten 100 ml Erlenmeyer flasks; two bored bungs; silicon tubing; graduated test tubes; Parafilm™.

Procedure

1 *Determining the kinetic constants of the enzymatic breakdown of H_2O_2 by soluble catalase.* 120 μl of catalase solution is diluted with 60 ml 0.01 M acetate buffer. 10 ml of the diluted catalase solution is placed in a 100 ml Erlenmeyer flask and then 100 ml of freshly prepared H_2O_2 solution is added. The flask is closed with the bored bung carrying the silicone tubing. One end of the tubing is placed in an upturned graduated test tube immersed in a beaker of water. The temperature of the reaction flask is brought up to 40 °C. The volume of water displaced by the oxygen released is measured in relation to time. 1 ml O_2 is equivalent to 45 μmol oxygen at room temperature. By plotting $(C_{S0} - C_S)/t$ against $(1/t)$ ln (C_{S0}/C_S), a straight line with the gradient tan $\alpha = -K_m$ and the point of intersection on the ordinate μ_{max} is obtained which corresponds to the integrated Michaelis–Menten equation

$$\frac{C_{S0} - C_S}{t} = -K_m \frac{1}{t} \ln \frac{C_S}{C_{S0}} + r_{max} \tag{3.87}$$

where r_{max} is compound as described in Eqn (3.65) and contains the amount of enzyme carrier used, c_E, in g/l.

2 *Immobilisation of catalase.* 10 g CM-cellulose of each of the particle sizes (0.2, 0.31, 0.4, 0.63 and 0.8 mm) is suspended in 200 ml 0.1 M acetate buffer, pH 4.5. After allowing the suspension to settle, the turbid supernatant is decanted and the cellulose beads are washed three times with 0.01 M acetate buffer, pH 4.5. The equilibrated ion exchanger is stirred for 45 min with 1 ml catalase diluted with 20 ml 0.01 M acetate buffer. The cellulose beads (complexed with catalase) are allowed to settle before decanting the supernatant.

3 *Determination of the activity of the immobilised enzyme.* 1 g each of the immobilized enzyme preparations are placed in 100 ml Erlenmeyer flasks with 10 ml 0.01 M acetate buffer. The flasks are covered with Parafilm™ and allowed to reach a temperature of 40 °C. Then 100 ml of dilute hydrogen peroxide solution (also at 40 °C) is added. The reaction is allowed to proceed with stirring, and the oxygen produced is measured with respect to time:

$$R_P = \frac{V_R}{w_K} \rho_K \frac{dC_P}{dt} \tag{3.88}$$

where R_P is the rate at which the product is formed, V_R is the volume of the reaction, w_K is the weight of the catalyser, ρ_K is the concentration of the catalyser and C_P is oxygen concentration.

Interpretation

Fig. 3.18 shows the curves obtained when the relative rate of mass change is plotted against time. R_{P0} is the rate of formation of the product within 5 min of starting the experiment. For the initial slopes at time $t \to 0$, the increase according to Eqn (3.78) is the product of γ_0 and k_d and should be calculated for each of the particle sizes.

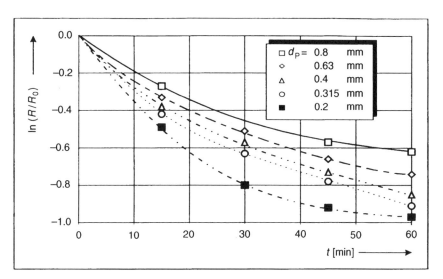

Figure 3.18 Dependence of the relative initial rates for catalysts of five different particle sizes on the reaction time.

Table 3.7 Documentation of the results of the experiment described in Subsection 3.2.4.

$\gamma_0 k_d$ [min^{-1}]	ϕ_0	γ_0	k_d [min^{-1}]	R_{P0} [mol/(m³ s)]	R_P [mol/(m³ s)]	η_0	η	θ

According to Eqn (3.69) the Thiele term is calculated from the data obtained with the soluble enzyme for r_{max} [mol/(l s)] and K_m [mol/l] and the effective diffusion constant $D_e = 2.2 \times 10^{-11}$ for H$_2$O$_2$ at 40 °C (Butt, 1980).

The inactivation factor γ_0 is calculated according to Eqn (3.80) using the Thiele term. For the preparation with the smallest particle diameter (0.2 mm) it can be assumed that $\eta_0 = 1$. From γ_0 and $\gamma_0 k_d$ the real deactivation constant k_d can be calculated. From Eqn (3.86) the relative half life θ for the immobilised enzymes can be determined as a function of the radius of the particles. The results are summarised in Table 3.7.

References

BAERNS, M., HOFMANN, M. and RENKEN, A. (1987) *Chemische Reaktionstechnik* (Stuttgart: Georg Thieme).

BUCHHOLZ, K. (1979) Characterization of immobilized biocatalysts. *Dechema Monographs* **84**.

BUTT, J.B. (1980) *Reaction kinetics and reactor design* (Englewood Cliffs, New Jersey: Prentice-Hall).

OSHIMA, H. and HARANO, Y. (1980) Effect of intraparticle diffusion resistance on apparent stability of immobilized enzymes. *Biotechnol. Bioeng.* **23**, 1991–2000.

3.2.5 *Determination of the Kinetics of Penicillin G Fermentation*

H. Voß

Principles and Aims

Penicillins are a class of antibiotics produced by strains of *Penicillium notatum* and *Penicillium chrysogenum* or obtained by partially synthetic methods. They all have a basic structure derived from 6-aminopenicillanic acid linked via an amide bond to various side chains. The activity of the penicillins was defined originally as the penicillin concentration which under given conditions just prevented the growth of a tester strain of *Staphylococcus aureus*. The international unit (iu) thus defined is equivalent to the activity of 0.6 μg 98% pure sodium salt of Penicillin G. Penicillin G is a relatively strong acid with a pK value of 2.75. It can be extracted from an aqueous solution at pH 2–3 (acidified with H_2SO_4) by organic solvents (amyl or butyl acetate). The sodium and other salts of Penicillin G are insoluble in the above-mentioned organic solvents but easily soluble in H_2O, methanol and ethanol. These differences in the solubility of the salts and the free acid of penicillin provide the basis for their isolation and purification.

The isolation of strains producing high levels of penicillin and the technology of penicillin fermentation are described in detail in Crueger and Crueger (1989).

Kinetics of the Process

The essential components of Penicillin G production are glucose, oxygen, CO_2, H_2O, NH_3, sulphuric acid, orthophosphoric acid, phenylacetic acid, mycelia, penicillin as the main product and penicillanic acid as by-product.

The changes in the concentration of substances during a discontinuous fermentation may be described by the following kinetic model.

Kinetics of Mycelial Growth

$$\frac{dX}{dt} = \mu X = \mu_{max} \frac{C_S}{K_S + C_S} X = r_X \tag{3.89}$$

Substrate utilisation for growth, endogenous metabolism and penicillin formation:

$$\frac{dC_S}{dt} = -\left[\left(\frac{1}{Y_{XS}}\right) r_X + m_{SX} + \left(\frac{1}{Y_{PS}} r_P\right)\right] = -r_S \tag{3.90}$$

Kinetics of Product Formation

There is a direct linear correlation observed between the growth rate μ and the rate of product formation r_P as long as $\mu < 0.01$ h^{-1}. If this value is exceeded then r_P is constant (Roels, 1983).

$$r_P = A' \frac{\mu}{0.01} X; \mu < 0.01 \text{ h}^{-1} \tag{3.91}$$

$$r_P = A'X; \mu \geq 0.01 \text{ h}^{-1} \tag{3.92}$$

In the experiment described here, only the first equation will be considered. Therefore

$$\frac{dC_P}{dt} = r_P \tag{3.93}$$

The rate at which the precursor is used up, the alkaline hydrolysis of penicillin and the enzymatic autolysis of the mycelia can be ignored.

Eqns (3.89) to (3.91) and (3.93) allow the process to be followed with sufficient accuracy once the constants have been determined.

First the yield coefficient is determined:

$$Y_{X/S} = \frac{\Delta X}{(-\Delta C_S)_X} \tag{3.94}$$

in the initial stage of growth, as well as the coefficient of endogeneous metabolism m_{SX} [(g glucose/g biomass) h], by plotting r_S/X against μ, where the slope of the straight line is $Y_{X/S}$ and the point of intersection on the ordinate is m_{SX}. The specific growth rate μ is calculated according to the equation

$$\mu = \frac{1}{X} \frac{\Delta X}{\Delta t} \tag{3.95}$$

The coefficient of penicillin formation can be calculated stoichiometrically and is approximately:

$$Y_{P/S} = 0.46 \text{ mol Penicillin G / mol glucose}$$

(Heijnen *et al.*, 1979). By plotting $t/[\ln(X/X_0)]$ against $\ln(C_{S0}/C_S)/[\ln(X/X_0)]$ one obtains a straight line with the slope A/μ_{max} and an intersection with the ordinate $(A + 1)/\mu_{max}$.

This method for calculating constants is known as the *integral method* because it depends on the integrated mass equilibrium of the batch fermenter with Monod growth kinetics (Fig. 3.19):

$$\mu_{max} t = \left(\frac{K_S}{C_{S0} + X_0/Y_{X/S}} + 1 \right) \ln \frac{X}{X_0} - \left(\frac{K_S}{C_{S0} + X_0/Y_{X/S}} \right) \ln \frac{C_S}{C_{S0}} \tag{3.96}$$

where

$$X - X_0 = Y_{X/S}(C_{S0} - C_S) \tag{3.97}$$

With the expression for A:

$$A = \frac{K_S}{C_{S0} + X_0/Y_{X/S}} \tag{3.98}$$

one obtains from Eqn (3.96) the linear form:

$$\frac{t}{\ln(X/X_0)} = \frac{A+1}{\mu_{max}} + \frac{A}{\mu_{max}} \frac{\ln(C_{S0}/C_S)}{\ln(X/X_0)} \tag{3.99}$$

which is the basis of the above-mentioned graphical representation obtained by the integral method.

To determine coefficient A' in Eqn (3.91) r_P/X is plotted against $\mu/0.01$ for the reaction phase following the addition of the precursor. A' may be determined from the slope of the curve. To all intents and purposes the following is valid:

$$r_P = \frac{\Delta C_P}{\Delta t} \tag{3.100}$$

With the set of kinetic and stoichiometric constants calculated it is possible to predict the fermentation process (altered reaction conditions or variation of the type of process, e.g. fed-batch).

Aim of the Experiment

By means of a batch fermentation of *Penicillium chrysogenum* the kinetics of the process of Penicillin G production will be calculated. The stoichiometric and kinetic constants will be calculated.

Description of the Experiment

Principle

Conidiospores are obtained from an agar culture of *Penicillium chrysogenum* and used as the starting material for the fermentation under aerobic conditions in the stirred fermenter vessel. The main fermentation is carried out with phenyl acetate as the precursor. From measurements of dry weight, glucose and the product (this is done by means of an agar diffusion test) changes in their concentrations as a function of time are noted, and from these values the stoichiometry and kinetics of the process can be determined.

Strains and Equipment

Microorganism: *Penicillium chrysogenum* (NRRL 1951 B25 or DSM 1075) on an agar slope or plate of Sabourand dextrose agar (Oxoid No. CM 42). Fernbach flasks (1800 ml; available from Schott Glaswerke, Germany); Erlenmeyer flasks (500 ml); sterile disposable pipettes (5 ml, 1 ml and 0.01 ml); sterilised flasks containing 100 ml H_2O; inoculating needles; Bunsen burner; bacterial filter; petri dishes; test tubes; 5 l fermenter; filter funnel; balance and microscope.

Procedure

The experiment to be carried out is one of those included in the Biotechnology practical course at the Technical University in Aachen, Germany.

1 *Culture of Penicillium chrysogenum.* Cell material from the stock culture of *Penicillium chrysogenum* is streaked out onto sporulation agar (7.5 g glycerol; 12.5 g sugar beet molasses; 2.5 g corn steep liquor (50%); 0.05 g $MgSO_4 \times 7$ H_2O; 0.06 g KH_2PO_4; 5.0 g peptone; 4.0 g NaCl; 5.0 g $FeSO_4 \times 7$ H_2O; 4.0 mg $CuSO_4 \times 5$ H_2O; 12.5 g agar; dH_2O ad 1000 ml; sterilised at 121 °C for 15 min). After 7 days incubation at 28 °C the surface(s) of the plate(s) should be covered evenly with conidiospores. The conidiospores are washed from the plate with

sterile water to which a few drops of Tween 80 have been added. The aqueous suspension of conidiospores is added to 300 ml culture medium in a Fernbach flask: 5.0% glucose; 6.0% corn steep liquor (50%); 0.02% NaNO$_3$; 0.05% KH$_2$PO$_4$; 0.02% MgSO$_4 \times 7$ H$_2$O; 0.8% CaCO$_3$; sterilised at 121 °C for 15 min. This flask is incubated for 2 days at 28 °C, at which time 30 ml of the culture is withdrawn and used as the inoculum for a second Fernbach flask containing 300 ml culture medium. This second flask is then incubated at 28 °C for 2 days.

2   *Fermentation.* For the fermentation, 4.2 l of substrate medium is prepared. Substrate medium consists of 3.5% lactose; 0.1% glucose; 0.4% KH$_2$PO$_4$; 6.0% corn steep liquor (50%); 0.2% olive oil. This medium is placed in the fermenter and the whole is sterilised according to the manufacturer's instructions.

During the fermentation phenyl acetate is added. 10 g phenyl acetic acid is dissolved in 30 ml 2.45 N NaOH and sterilised in the dispensing pump. The first addition of 5 ml phenyl acetate is made 15 h after the start of fermentation and then in 5 ml portions every 24 h.

20 ml antifoam is sterilised separately, and 1% (with respect to the fermenter volume) CaCO$_3$ (powdered) is placed in a glass container, closed with a cotton wool bung and sterilised by placing in a drying oven at 160 °C for 3 h. When cool, the CaCO$_3$ is resuspended in 200 ml sterile H$_2$O and the suspension is added to the fermenter after inoculation.

Fermentation conditions: temperature 24 °C, stirring speed 600–800 rpm, aeration 1 vvm.

KINETIC MEASUREMENTS

1   At 24 h intervals 150 ml samples are removed from the fermenter and the following information is determined:

(i)   dry weight of the mycelium

(ii) pH value

(iii) penicillin content

(iv) microscopical examination for possible contamination.

2   To determine the dry weight of the mycelia 100 ml of the suspension is vacuum-filtered. The filtrate is to be examined further and is therefore removed before the mycelia are washed with 1% hydrochloric acid to remove the CaCO$_3$. The washed mycelia and the filter paper are dried at 105 °C for 5 h and, after cooling, are weighed. The dry weight of the filter is subtracted from the total weight.

3   Penicillin content is measured by a plate diffusion test. For this purpose 20 ml sterile Standard-I-agar is poured into plastic petri plates and allowed to dry over a period of 3 days before use. *Preparation of plates*: antibiotic medium No. 5 or Standard-I-agar; 6 g peptone; 3 g yeast extract; 1.5 g meat extract; 5 ml 1N NaOH; 10 g agar; dH$_2$O ad 1000 ml, pH 7.9.

4   50 ml Standard-I-nutrient broth is inoculated with the tester strain *Bacillus subtilis* ATCC 6633 and incubated, with shaking, for 3 h at 28 °C. 5 ml of the prepared Standard-I-agar is melted, cooled to 48 °C and inoculated with 0.1 ml of the culture. This bacteria-containing agar is spread over the surface of a 3-day-old Standard-I-agar plate and allowed to solidify for 1 h at room temperature. Using a sterile cork borer (diam. 8 mm), five wells are made in the agar.

The wells are filled with 0.2 ml of standard solutions containing 2.0, 1.5, 1.0, 0.5 and 0.25 iu/ml Penicillin G. Other petri dishes prepared in the same manner are inoculated with dilutions of the filtrate from step 2. The plates are then incubated at 28 °C for 16–18 h. During this time penicillin diffuses from the wells into the agar, inhibiting growth of the tester strain and causing clear zones of inhibition to appear around the wells. The diameters of the zones of inhibition are measured and provide a measure of the effectivity or concentration of the Penicillin G solution tested. A standard curve based on the measurements from 10–20 standard plates allows the iu/ml in the culture filtrates to be determined.

5 Concentration of Penicillin G can be determined by using the approximation that 1 iu/ml $\approx 0.6 \, \mu g/l$. It is preferable to draw up a standard curve in mg/l. The correlation between the log of the penicillin concentration in mg/l and the diameter (or the square of the radius) of the zone of inhibition is obtained.

6 The glucose and lactose concentrations in the fermenter medium can be determined enzymatically as described by Bergmeyer (1977) using lactose/glucose standard tests available from Boehringer Mannheim.

Interpretation

The concentration measurements are presented in tabular form and as graphs plotted against time. The glucose, lactose and penicillin concentrations and the dry weight of the mycelia are plotted as a function of time and the measured values. From the curves obtained, the kinetic constants – as described above – are calculated. Numerical methods for parameter optimisation (Müller-Erlwein, 1991) are also to be recommended.

Bibliography

BERGMEYER, H.U. (1977) *Grundlagen der enzymatischen Analyse* (Weinheim: VCH).

BIRDY, J. (1974) Recent developments of antibiotic research and classification of antibiotics according to chemical structure. *Adv. Appl. Microbiol.* **18**, 309–406.

CRUEGER, W. and CRUEGER, A. (1990) *Biotechnology: a textbook of industrial microbiology* (Sunderland, MA: Sinauer Associates, Inc.).

DREWS, G. (1983) *Mikrobiologisches Praktikum* (Heidelberg: Springer Verlag).

HEIJNEN, J.J., ROELS, J.A. and STOUTHAMER, A.H. (1979) Application of balancing methods in modelling the penicillin fermentation. *Biotechnol. Bioeng.* **21**, 2175–201.

MÜLLER-ERLWEIN, E. (1991) *Computeranwendungen in der Chemischen Reaktionstechnik* (Weinheim: VCH).

ROELS, J.A. (1983) *Energetics and kinetics in biotechnology* (Amsterdam: Elsevier).

3.2.6 *Energetics of Discontinuous and Semi-continuous Fermentation in Yeast*

H. Voß

Principles and Aims

The production of yeast is one of the classical processes in biotechnology but is nevertheless also an up-to-date research topic in bioprocessing.

Yeast biomass is produced not only for the baking industry but also for pharmaceutical and microbiological purposes and is used for animal feed in the farming industry. For the sake of easier handling, the initial stages of production are carried out either discontinuously or in batches. An example of such a production stage will be described below.

In the factory, yeast is produced in a so-called 'despatch fermenter', where the yeast grows under conditions ensuring the best possible use of space, time and the amount of yeast produced. Optimal provision of nutrients and monitoring of pH and temperature, as well as optimal aeration and other measures are essential. The possibilities for yeast growth are however restricted at the 'despatch stage' for a number of reasons. One cannot for instance take a normal laboratory culture to inoculate the production fermenter, which has a volume of 100 m^3. The volume of yeast has to be enlarged steadily through several fermenters of increasing size. A sign of quality of the yeast is its 'raising' power.

Physiology

Saccharomyces cerevisiae can grow both aerobically (using atmospheric oxygen) and anaerobically (in the absence of oxygen). In the presence of oxygen the hydrogen produced by substrate metabolism is transferred to intermediate metabolites thereby reducing them and releasing them as fermentation products (e.g. acetaldehyde $\rightarrow$ ethanol).

S. cerevisiae metabolism is characterised by the fact that when sufficient oxygen is available, glucose at a concentration in excess of 0.02% can be partially fermented to ethanol and CO_2, resulting in a reduction in the amount of biomass obtained. This phenomenon, peculiar to baker's yeast, is known as the *Crabtree effect*. Simultaneously the enzymes of oxidative glycolysis and the citric acid cycle are inhibited. At a glucose content of 0.1% or more, even when sufficient dissolved oxygen is available, it is still glucose which is fermented.

Kinetics

At substrate concentrations which are not inhibitory ($C_S < 50$ g/l) the growth kinetics of baker's yeast in a batch fermenter can be described by the following kinetic model. Biomass:

$$\frac{dX}{dt} = \mu X \tag{3.101}$$

at the growth rate μ (h^{-1}) according to Monod:

$$\mu = \mu_{max} \frac{C_S}{K_S + C_S} \tag{3.102}$$

where μ_{max} is the maximum growth rate (h^{-1}) and K_S (g/l) is the saturation constant. The breakdown of the substrate can be expressed as the following equilibrium:

$$\frac{dC_S}{dt} = -\frac{1}{Y_{X/S}} \mu X \tag{3.103}$$

with the yield of the biomass $Y_{X/S}$ expressed in terms of the substrate as:

$$Y_{X/S} = \frac{X - X_0}{C_{S0} - C_S} \tag{3.104}$$

Integration of Eqn (3.101) with μ remaining constant gives

$$X = X_0\, e^{-\mu t} \tag{3.105}$$

Therefore the doubling time t_d for $X = 2X_0$ is

$$t_d = \frac{\ln 2}{\mu} \tag{3.106}$$

The precise time course of the concentrations $X(t)$ and $C_S(t)$ is obtained when Eqn (3.101) is expressed in terms of Eqns (3.102) and (3.104):

$$\frac{dX}{dt} = \mu_{max} \frac{C_{S0} - \dfrac{1}{Y_{X/S}}(X - X_0)}{K_S\left[C_{S0} - \dfrac{1}{Y_{X/S}}(X - X_0)\right]} X \tag{3.107}$$

Integration gives:

$$\mu_{max}\, t = \left(\frac{K_S}{C_{S0} + \dfrac{X_0}{Y_{X/S}}} + 1\right)\ln\frac{X}{X_0} - \left(\frac{K_S}{C_{S0} + \dfrac{X_0}{Y_{X/S}}}\right)\ln\frac{C_S}{C_{S0}} \tag{3.108}$$

The time course of $C_S = f(t)$ can be determined from Eqn (3.108), substituting Eqn (3.104) for X:

$$X = X_0 + Y_{X/S}(C_{S0} - C_S) \tag{3.109}$$

Eqn (3.108) can be rearranged to give

$$\frac{t}{\ln(X/X_0)} = \frac{A+1}{\mu_{max}} + \frac{A}{\mu_{max}}\frac{\ln(C_{S0}/C_S)}{\ln(X/X_0)} \tag{3.110}$$

allowing $(A + 1)/\mu_{max}$ to be obtained from the intersection with the ordinate, as in Fig. 3.19; and since the slope is A/μ_{max} and

$$A = \frac{K_S}{C_{S0} + \dfrac{X_0}{Y_{X/S}}} \tag{3.111}$$

the constants μ_{max} and K_S can be determined directly from the values measured.

With the help of batch kinetics other processes such as continuous or semi-continuous fermentations can be predicted. It should of course be remembered that the kinetics can be affected by changes in the culture conditions.

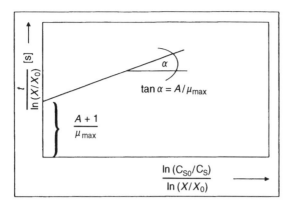

Figure 3.19 Calculation of the kinetic constants from the Monod model using an integral method, $A = K_s/[C_{s0} + (X_0/Y_{x/s})]$.

Fed-batch Fermentation

Fed-batch fermentation is a semi-continuous fermentation in which only a small amount of substrate is available at the beginning of the process. During the process, substrate is added as required by the growth rate or the rate at which the product is formed (Fig. 3.20). The specific rate at which the substrate is used up, $q_{s/x}$, can be determined as the amount of substrate per cell and unit of time from the growth kinetics:

$$q_{x/s} = \frac{1}{Y_{x/s}}\,\mu \tag{3.112}$$

The specific rate of substrate uptake $q_{s/x}$ (g/l h^{-1}) in a fed-batch culture is satisfied by addition on demand. The required volumetric feeding rate, $Q_s(t)$, consists of $q_{s/x}$ and the cell density $X(t)$:

$$Q_s(t) = q_{s/x}\,X(t) \tag{3.113}$$

This must be identical to the feeding rate:

$$Q_s(t) = \frac{F(t)C_{s0}}{V(t)} \tag{3.114}$$

where $F(t)$ is the rate of pumping (l/h) at the given time, C_{s0} is the concentration of the input and $V(t)$ is the volume of the reaction. The method of feeding can be either constant, which results in linear growth, or adjusted to increase exponentially so that C_s is maintained at an optimal level and results in exponential growth. The balance in a fed-batch fermenter may be described as follows. Biomass:

$$\frac{d(VX)}{dt} = \mu XV \tag{3.115}$$

from which

$$\frac{dX}{dt} = (\mu - D_f)X \tag{3.116}$$

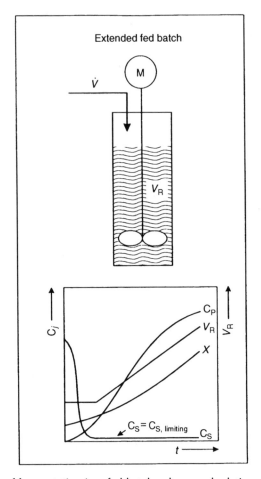

Figure 3.20 Course of fermentation in a fed-batch culture and relative concentrations during the fermentation.

Since the volume increase as a result of the input is

$$\frac{dV}{dt} = F \tag{3.117}$$

D (the 'dilution rate' as a result of input) is

$$D = \frac{F}{V} \tag{3.118}$$

For limiting substrate the following is valid:

$$\frac{d(VC_S)}{dt} = C_{S0} F - \left(\frac{\mu X V}{Y_{X/S}}\right) \tag{3.119}$$

$$\frac{dC_S}{dt} = (C_{S0} - C_S)D - \left(\frac{\mu X}{Y_{X/S}}\right) \tag{3.120}$$

The equilibrium equations (3.116) and (3.120) can be solved analytically under certain conditions (Keller and Dunn, 1987) so that the optimal strategy for substrate input to achieve maximum amounts of product (biomass) can be calculated.

For the present situation the optimal production of baker's yeast is particularly interesting. The substrate concentration in the fermenter should be kept at a constant level to prevent the Crabtree effect but nevertheless to achieve an optimal production of yeast. Therefore the substrate concentration during the experiment must be regulated. Eqn (3.115) is integrated and, when μ remains constant,

$$X = \left(\frac{V_0 X_0}{V}\right) e^{\mu t} \tag{3.121}$$

From Eqn (3.119) it follows when C_S is constant that:

$$F = \frac{XV}{Y_{X/S}(C_{S0} C_S)} \tag{3.122}$$

By combining Eqn (3.122) with Eqn (3.121) the growth-dependent exponential feeding rate for constant C_S in the fermenter is given by Eqn (3.123). The value of μ depends on the desired value for C_S.

$$F(t) = \frac{X_0 V_0}{Y_{X/S}(C_{S0} - C_S)} e^{\mu t} \tag{3.123}$$

In the fermenter the volume increases exponentially as follows:

$$V(t) = V_0 + \int_0^t F \, dt \tag{3.124}$$

$$V(t) = V_0 + \frac{X_0 V_0}{Y_{X/S}(C_{S0} - C_S)} (e^{\mu t} - 1) \tag{3.125}$$

from which the concentration of biomass X in the fermenter at time t can be calculated by substituting Eqn (3.125) in Eqn (3.121):

$$X(t) = \frac{X_0}{e^{-\mu t}\left[1 - \dfrac{X_0}{Y_{X/S}(C_{S0} - C_S)}\right] + \left[\dfrac{X_0}{Y_{X/S}(C_{S0} - C_S)}\right]} \tag{3.126}$$

Specific Aim of the Experiment

1   The growth kinetics according to Monod and the coefficients μ_{max}, K_S and $Y_{X/S}$ should be calculated from the discontinuous yeast fermentation.

2   Using the information gained above, a fed-batch experiment should be carried out with optimal (exponential) feeding of the substrate, and the values obtained for $X = f(t)$ are to be compared with the calculations of Eqns (3.121) and (3.126).

Description of the Experiment

Principle

Throughout the aerobic batch fermentation the biomass and the substrate concentrations are measured against time, and the results are presented graphically. Using the graphically 'corrected' values the Monod constants can be calculated by the integral method.

The yield coefficient $Y_{X/S}$ is calculated from the linear regions of the substrate utilisation and biomass production curves.

$$Y_{X/S} = \frac{\Delta X}{-\Delta C_S} \tag{3.127}$$

Finally, an aerobic fed-batch fermentation is carried out in which the rate of feeding will be controlled by the growth rate μ calculated for a specific constant substrate concentration.

Material

30 l laboratory fermenter with a working volume of 25 l, fitted with pH and pO_2 measuring devices; 1 l, 2 l and 10 l flasks for the increasing volumes of inoculum; sampling vessels; test tubes; eight 25 ml volumetric flasks; 25 ml Erlenmeyer flask; 1 ml, 2 ml, 10 ml and 20 ml pipettes; 1 l flask for medium; inoculating flask containing 100 ml water and a magnetic stirring bar (sterilised); sterile syringe for NaOH; flask for NaOH; test tubes (50); three sample tubes, one of which should be sterile; 15 YEP plates (see Subsection 2.3.7); test tube rack; cuvettes; ten 25 ml flasks; ten centrifuge tubes; centrifuge; water bath with test tube rack; five small magnetic stirring bars; vortex mixer; magnetic stirrer; spatula.

Reagents and organism: enzymatic glucose test kit (Boehringer Mannheim); 600 g commercially available baker's yeast.

Media: Media for batch kinetics; medium for fed-batch experiment and feeding medium for the fermenter. All media are described in Table 3.8.

Inoculum:

- For the batch kinetic, 11.32 g yeast is resuspended in 100 ml sterile tap water.
- For the fed-batch kinetic, 85 g of yeast (approx. 27% dry weight of yeast) is resuspended in 0.6 l sterile tap water and then added to the sterilised medium already in the fermenter to give $X_0 = 5$ g/l.

Fermentation Conditions and Other Requirements

1 The stirring mechanism available in the fermenter reaches an oxygen transfer rate of 2050 ml $O_2/(l\ h)$ under the given conditions (rate of aeration = 0.05 vvm; speed of the stirrer = 500 rpm; $T = 30\,°C$).

2 Since 520 ml O_2 is required to produce 1 g dried yeast, the maximum amount of dried yeast which can be produced per hour and litre is 3.95 g or 14.6 g yeast with 27% dry weight. With a working volume of 25 l this amounts to the production of 98.75 g dried yeast or 365.7 g yeast per hour in the fermenter. This increase in the amount of yeast at a yield constant of 0.54 requires sucrose to be fed at the rate of ... g/h. Complete use of the fermenter will be achieved if the

feeding during the log phase is allowed to reach this value and the value is maintained in the following linear phase. If the feeding rate is exceeded there will be a loss in the yield as a result of alcohol formation.

The feeding scheme and the expected values can be seen in Tables 3.8 and 3.9.

Procedure

The fermenter is sterilised the evening before the experiment is to be carried out. The container for the pH monitor is filled with 2N NaOH.

The stirrer is switched on about 15 min before starting the fermentation; the temperature (30 °C), the pH value (4.8) and the aeration (0.5 vvm) are adjusted and the input dosage system is set up.

Table 3.8 Composition of medium for propagation of yeast cells.

Component	g/l^a	Amount weighed in [b]	Medium for fed-batch experiment[c]	Added medium containing 90% minerals[d]	
Glucose	11.23	45			
$(NH_4)_2SO_4$	2.25	9	4.2 g	143	g
$(NH_4)_2HPO_4$	0.72	2.9	0.8 g	23	g
KCl	0.326	1.3	0.8 g	17	g
$MgSO_4 \times 7 H_2O$	0.17	0.68	0.4 g	10	g
$CaCl_2 \times 7 H_2O$	0.17	0.68			
Yeast extract	0.5	2	0.5 g		
Tap water (source of trace elements)		4.3 l			
Yeast			85 g	13.3 g	
Sugar				2000	g
Tap water				20	l

[a, b] Composition and amounts required for the batch kinetic.
[c] Medium for fed-batch experiments. The medium in the reservoir (V_0) contains about 10% of the amount of minerals and trace elements. The amounts given are dissolved in 4.0 l tap water.
[d] 90% of the minerals are contained in the medium to be fed into the fermenter.

Table 3.9 Expected values for the yeast fermentation.

Growth phase	Time (h)	pH	Aeration (vvm)	h^{-1}
lag	1	1.13	4.8	0.3
log	2	1.17	4.8	0.4
log	3	1.17	4.9	0.4
log	4	1.17	5.0	0.5
log	5	1.17	5.1	0.5
lin	6	?	5.2	0.5

The batch fermentation takes about 6–12 h. 10 ml samples are withdrawn every hour and the biomass and the substrate concentration (determined according to the manufacturer's instructions in the test kit) are measured.

The biomass can be determined gravimetrically after filtering and washing the sample or by measuring the OD at 600 nm. The intensity of the light entering (I_0) and that passing through the suspension (I_t) is related to the cell density (x) and the light path (L) of the cuvette (width of the cuvette 1 cm) by the Lambert–Beer law:

$$\log(I_0/I) = Axl$$

The expression $\log(I_0/I)$ is known as optical density (OD) or extinction. The factor A is constant for lower numbers of microorganisms, but decreases at higher cell concentrations because of light scattering in the suspension. If necessary the suspension must be diluted. With the aid of a calibration curve which reflects the dependency of the value measured on the number of particles (cells) per unit volume, the results of the turbidity measurement can be used to determine the cell concentration. The calibration curve is obtained by weighing the relative amounts of baker's yeast. For the fed-batch experiment 4 l of the fermentation medium is inoculated with the yeast suspension. After 10 min a 10 ml sample is taken and X_0 is determined analytically.

A dosage programme $F(t)$ according to Eqn (3.123), substituting X_0 (5 g/l), V_0 (4.6 l) and C_{S0} (10 g/l), the required C_S in the fermenter of 1 g/l and μ as determined previously in the batch kinetic, is calculated and checked half-hourly at the input pump. By calculating the increase in volume with respect to time (Eqn (3.125)) it is possible to determine the time the experiment will take and the amount of biomass produced at the end of the experiment. When the samples are removed hourly compressed air is blown through the sample tube which is then flamed before allowing about 50 ml of the suspension to run into a collecting vessel. A 10 ml sample is then withdrawn into the sampling vessel. The sampling tube is again flushed with compressed air.

The samples are used immediately to determine the amount of yeast dry weight and the glucose concentration.

In addition to the hourly samples the following should be determined every 30 min:

- the rate of aeration
- the speed of the stirrer
- the amount of NaOH used
- the input dosage
- pH value

and a table drawn up for comparison of the expected and observed values.

In order to expand Table 3.10 to an input plan the total amount of sucrose added in the fifth hour should be calculated on the assumption that at the end of the fifth hour the maximum rate of ... g/h is reached.

- Sucrose input during the fifth hour: ... g; therefore:
- Increase in the amount of yeast in the same period = A_5: ... g H_{27}. During the log phase the following conditions are valid between Z_0 (the amount of yeast at the start of the fermentation period t) and Z_h (the hourly increase in the amount of yeast at the particular hour t):

$$Z_0 = A_0(H^t - 1)$$
$$Z_h = A_0 H^{t-1}(H - 1)$$

Table 3.10 Input plan for the fermentation (should be completed before the start of the experiment).

Time (h)	H value (h^{-1})	Yeast (total as H_{27}) (g)	Yeast (increase in H_{27}) (g/h)	Sucrose input (g/h) (ml/h)
0		$A_0=$	$Z_h=$	
1	1.13			
2	1.17			
3	1.17			
4	1.17			
5	1.17			

Based on these equations A_0 can be calculated starting from A_5 and the input plan can be expanded. (NB: the fifth hour of the fermentation is the fourth hour of the log phase.)

Thus, the amount of yeast required at the start of the fermentation can be determined. The yeast is added as an aqueous suspension in 2 l of sterile tap water. At the beginning of the fermentation the volume of the fermentation culture should be 25 l, made up of 23 l of medium and 2 l of the yeast suspension. At the end of the sixth hour of the fermentation the fermenter is connected directly to a separator so that the yeast can be removed from the spent medium. The yeast is then washed. The dry weight of the resulting yeast suspension is determined as described.

Interpretation

1 Using the terms $X = f(t)$ and $C_S = f(t)$ obtained from the batch experiment, first of all $Y_{X/S}$ and then, by integration, μ_{max} and K_m are to be calculated.

2 The expected and the observed values for X, C_S, $F(t)$ and $v(t)$ should be compared.

3 Furthermore the productivity of the fed-batch fermenter $[g/(l\,h^{-1})]$ must be determined.

By plotting X against t a curve is obtained whose linear region is characterised by the increase $F\,C_{S0}\,Y_{X/S}$ from which the yield factor can be determined and compared with $Y_{X/S}$ from the batch experiments.

Further reading

BOHL, E. (1987) *Mathematische Grundlagen für die Modellierung biologischer Vorgänge* (Heidelberg: Springer Verlag).

KELLER, R. and DUNN, I. (1987) Fed-batch microbial culture: models, errors and applications. *J. Appl. Chem. Biotechnol.* **28**, 508–14.

MCNEIL, B. and HARVEY, L.M. (1990) *Fermentation – a practical approach* (Oxford: IRL Press).

MOSER, A. (1981) *Bioprozeßtechnik* (Heidelberg: Springer Verlag).

3.2.7   *Kinetics of Enzyme Reactions: Reduction of NAD $^+$ to NADH by Formiate Dehydrogenase and Formic Acid*

H.-P. Schmauder

Thanks are due to C. Wandrey, who developed the following experiment and tested it in practical courses at the University of Bonn, for allowing it to be used in this book.

A simplified kinetic equation describing an enzymatic reaction with two substrates is multiplication of two Michaelis–Menten kinetics for single-substrate kinetics as shown in the equation of the reaction:

$$A + B \rightarrow P$$

An example of such a reaction is the reduction of NAD^+ to NADH by formiate dehydrogenase (FDH) and formic acid:

$$NAD^+ + HCOO^- \xrightarrow{\text{FDH}} NADH + CO_2$$

The equilibrium of this reaction lies totally on the side of the product and it can be used for the regeneration of the cofactor by the reduction of carbonyl groups, where the cofactor is used either in its native state or enlarged to a polymer (Kragl *et al.*, 1992).

The kinetics are described in Eqn (3.128).

$$V = A_{\text{mass}}[E] = [E]A_{\text{max}} \frac{[A]}{K_m^A(1 + [P]/K_i^P) + [A]} \frac{[B]}{K_m^B + [B]} \tag{3.128}$$

where:

- V (U/ml) is reaction rate of the synthesis
- A_{max} (U/mg) is maximum specific activity of the synthesis
- A_{mass} (U/mg) is specific activity of the synthesis
- $K_m^{A,B}$ (mol/l) are Michaelis–Menten constants of A and B
- K_i^P (mmol/l) is inhibition constant of P, a competitive inhibitor
- $[E]$ (mg/ml) is concentration of the enzyme
- $[\,]$ (mmol/l) are concentrations of A, B and P.

The unit of enzyme activity is defined as:

$$1 \text{ U} = 1 \ \mu\text{mol/min}$$

The SI unit is cat:

$$1 \text{ cat} = 1 \text{ mol/s}$$

In this example it will be demonstrated how the kinetic parameters of a reaction can be determined by measuring the initial reaction rate. The increase in OD caused by the formation of NADH is measured at 340 nm. NAD^+ and the other components of the reaction do not absorb at this wavelength.

By varying the substrate concentration and determining the corresponding initial reaction rate the kinetic parameters can be obtained from the rate/concentration curve by various methods: linearisation and linear regression, non-linear regression (Cornish-Bowden, 1981).

In the same reaction system it will be demonstrated how inhibition by the reaction products influences the rate of the enzyme reaction. For this purpose rate/concentration curves at different product concentrations will be measured and the constants of inhibition determined. There are technical methods for reducing the product inhibition:

- choice of the appropriate reactor (flow cylinder, modular reactor, stirred vessel cascade)
- selective removal of the product (electrodialysis, chromatography) or
- incorporation of a follow-up reaction

to name but a few possibilities.

Conditions

- 25 °C, 50 mM phosphate buffer, pH 8.0
- wavelength 340 nm, path length 1 cm, 0.1 cm, 0.02 cm
- $[NAD^+]$ 0–1 mM
- $[HCOOH]$ 0–200 mM (as formiate because of the pH value)
- $[NADH]$ 0–1 mM
- $[FDH]$ ca. 20 $\mu g/ml$

Procedure

The cuvette and the reaction mixture (without the enzyme!) are prewarmed at 25 °C for about 3 min. The reaction is started by the addition of enzyme. The cuvette is placed in the photometer and the increase in OD_{340} is measured over a period of 3 min (ideally the photometer should have a thermostatically controlled cuvette holder).

The specific activity of the enzyme under the chosen conditions is calculated from the increase in OD_{340} with respect to time according to Eqn (3.129).

$$A_{mass} = \frac{(dA/dt)\, dil^n}{d\, 6.220[E]} \tag{3.129}$$

where:

- A_{mass} (U/mg) is specific activity of the enzyme
- dA/dt (1/min) is change in OD with respect to time
- dil^n is dilution of the enzyme stock solution used in the assay
- d (cm) is path length of the cuvette
- 6.22 [ml/(μmol cm)] is coefficient of extinction of NADH at 340 nm
- $[E]$ (mg/ml) is enzyme concentration in the stock solution.

Solutions required: phosphate buffer 50 mM, pH 8.0; NAD^+ 0.5 mM and 5 mM in buffer (check pH!); NADH 1 mM in buffer (check pH!); $HCOONH_4$ 1 M in buffer; FDH 5 mg/ml, made up in buffer and diluted 1 in 20.

Table 3.11 Turnover of NAD regeneration – pipetting scheme.

[NAD$^+$] (μM)	NAD$^+$ sol^n (μl)	Formiate sol^n (μl)	NADH sol^n (μl)	Buffer (μl)[a]	FDH (μl)	Activity (U/mg)
0	0	200	0/20/100	750/730/650	50	
2.5	5	200	0/20/100	745/725/645	50	
5	10	200	0/20/100	740/720/640	50	
10	20	200	0/20/100	730/710/630	50	
20	40	200	0/20/100	710/690/620	50	
40	80	200	0/20/100	670/650/570	50	
100	20[b]	200	0/20/100	730/710/630	50	
150	30[b]	200	0/20/100	720/700/620	50	
200	40[b]	200	0/20/100	710/690/610	50	
250	50[b]	200	0/20/100	700/680/600	50	

[a] additions at [NADH] concentrations of 0, 20 and 100 μM.
[b] concentrated [NADH] solution.

REACTION MIX

This is a variation of NAD$^+$ concentration, Table 3.11.

- 200 μl formiate solution →200 mM in the reaction mix
- 0–100 μl NADH solution →0–100 mM in the reaction mix
- 0–80 μl NAD$^+$ solution →0–250 mM in the reaction mix
- max 750 μl buffer
- 50 μl FDH solution →12.5 μg in the reaction mix
- 1000 μl total volume.

The volume of the buffer added depends on the volume of the substrate solution.

Interpretation

The enzyme activity is calculated at different substrate concentrations and its dependency thereon is calculated using Eqn (3.128).

References

CORNISH-BOWDEN, A (1981) *Fundamentals of enzyme kinetics* (London: Butterworths).
KRAGL, U., VASIC-RACKI, D. and WANDREY, C. (1992) Continuous processes with soluble enzymes. *Chem.-Ing.-Tech.* **64**, 499–509.

3.3 IMMOBILISATION OF CELLS FOR BIOTRANSFORMATION

D. Schlosser and H.-P. Schmauder

Principles and Aims

Biotransformations are chemical reactions carried out on natural or synthetic products by biocatalysts (enzymes, cells, etc.) and which have several advantages over chemical processes: e.g.

- high specificity of reactions with only few side products
- very often, selection of stereospecificity, depending on the biocatalysts used and the substrates
- strict regiospecificity, depending on the biocatalysts used and the substrates
- mild reaction conditions offering protection to substrates and products.

Biotransformations can be carried out with different techniques:

- conversion in growing cultures after addition of the substrate to be converted at the beginning of the culture (sterility is absolutely essential!)
- conversion in fully grown cultures when the substrate is added to the cells after growth has ceased or to a concentrated suspension of biologically active cells
- conversion using isolated, more or less purified enzymes
- conversions with immobilised cells or enzymes
- biotransformations in liquid two-phase systems; such systems are useful for lipophilic substances and when the biocatalysts are insensitive to organic, water-immiscible solvents.

When using cells as biocatalysts it is essential that the substrate and product can move freely between the phases. Processes in which biotransformation can be used are:

- formation of C−C bonds
- insertion of chiral centres
- separation of racemates
- regioselection of functional groups in the neighbourhood of equivalent or other active groups, without having to introduce protective groups
- regioselective activation of inactive C atoms
- specific *de novo* synthesis of natural products using modified substrates, conversions and partial syntheses, e.g. penicillin, mutasyntheses.

If the regeneration of cofactors is necessary for the processes or if multienzyme complexes are required, cells are the preferred biocatalysts since coupling systems for cofactor regeneration perform as enzymes only with difficulty under optimal conditions (see Subsection 3.2.7).

Immobilization is the technique of binding biocatalysts to a carrier as a means of increasing their activity and stability, improving the technological application of the reaction (e.g. separation of the products from the biocatalyst; increasing the number of times a reaction can be performed).

Possible means of immobilisation are:

- binding to inorganic or other polymer carriers (glass, silicates, polystyrenes, etc.)
- enclosure in organic, gel-forming matrices (e.g. Ca alginate, agar, carragenane, PAA)
- enclosure in light-sensitive polymers (e.g. derivatives of polyvinyl alcohol)
- adsorption or another means of fixation to or in membranes (e.g. for membrane reactors).

Examples of the various methods will be demonstrated in the experiment. The model reaction chosen is the long established *11β-hydroxylation of Reichstein's substance S (RSS)* by *Curvularia lunata*. This hydroxylation is used in many industrially viable syntheses of steroid active substances. The second reaction is the 15 α-hydroxylation of *13-ethyl-gon-4-en-3,17-dion (I)*, carried out by the imperfect fungus *Penicillium raistrickii* 477. The use of cyclodextrins illustrates the use of this class of compounds to improve the solubility of lipophilic substrates or products.

EXPERIMENTAL PROCEDURE

Strain Cultivation

The strains are grown on slants of malt extract agar (malt extract 4.0%, yeast extract 0.3%, agar 1.5%, made up in dH_2O, pH 6.6–6.8, sterilised at 120 °C for 20 min).

The slants are incubated at 28 °C for 7 days; on this medium *Curvularia lunata*, in contrast to *Penicillium raistrickii*, does not form spores.

Preparation of Spore Suspensions

1 *Curvularia lunata*. Slants of oatmeal agar (oatmeal 2%, agar 2%, made up in dH_2O, pH 4.7–4.9, sterilised at 120 °C for 20 min) are incubated at 28 °C for 7–14 days. The spores are washed from the agar surface by two 10 ml aliquots of 0.9% NaCl. The spore suspension is filtered through sterile cotton wool or glass wool placed in a filter funnel, to remove any fragments of hyphal tissue. The spore suspension can then be immobilised. (*Caution!* The spores germinate particularly well at 37 °C. Inhalation of spore-containing aerosols or dust should be avoided.)

2 *Penicillium raistrickii*. As described above but using malt extract agar instead of oatmeal agar.

Biotransformation

The two reactions named above will be carried out as biotransformations with growing cells. 50 ml of the following medium (glucose 30.0 g, corn steep liquor 10.0 g, $NaNO_3$ 2.0 g, KH_2PO_4 1.0 g, K_2HPO_4 2.0 g, $MgSO_4 \times 7 H_2O$ 0.5 g, $FeSO_4 \times 7 H_2O$ 0.02 g, KCl 0.5 g, dH_2O ad 1000 ml; sterilised at 120 °C for 20 min) are placed in 500 ml round-bottom flasks. The flasks are inoculated with either an immobilised culture of spores or a suspension culture of the same and incubated at 28 °C with shaking (180 rpm).

6–24 h later, methanolic solutions of Reichstein's substance S (RSS) or 13-ethyl-gon-4-en-3,17-dion are added aseptically (100 mg/50 ml, 2 vol% methanol) and the biotransformation is monitored over a period of several days.

The biotransformation can also be carried out with resting cells. For this variation of the method 24-h-old cultures of immobilised cells or free cells are resuspended in 0.9% NaCl, and 50 ml aliquots are placed in the flasks. RSS or 13-ethyl-gon-4-en-3,17-dion are added immediately.

N.B. When cells are immobilised with Ca alginate, all solutions used must contain 0.01–0.05 M $CaCl_2$.

Analysis of Steroids

On the first, second and third days, 3 ml samples are withdrawn and shaken with 2 vols of ethyl acetate for 1 min. 200 μl of the upper organic phase is transferred to Eppendorf reaction tubes and the solvent allowed to evaporate. The residue is taken up in 1 ml acetonitrile and diluted 1:5 in acetonitrile. The diluted sample is then used for HPLC analysis. Detection 242 nm; liquid phase: acetonitrile/water = 70/30; injection 20 μl; column RP-8 (250 × 4 mm); flow rate 1.5 ml/min.

Immobilisation in Ca Alginate Gel Beads

1   2 g sodium alginate is dissolved in 100 ml 0.9% NaCl; if necessary the solubilisation can be speeded up by warming the solution gently, on no account should it be allowed to boil! The solution is then autoclaved for no longer than 15 min.

2   10 ml of a spore suspension is added to the solution prepared in step 1.

3   The spore/Na alginate solution is then added dropwise (e.g. with a burette) with stirring to a 0.1 M CaCl$_2$ solution. The gel beads formed are left in solution for 1 h before being filtered off. The beads are then stirred in a 0.9% NaCl / 50 mM CaCl$_2$ solution for 20 min to allow the diffusion of excess calcium. The washed, cell-coated beads are now ready for use.

Immobilisation in Photolinkable Prepolymers

One type of photolinkable prepolymer is polyvinyl alcohol, acetylated with stilbazolium salts as the photoreactive groups (0.228 mmol stilbazolium per gram polyvinyl alcohol). The immobilisation can be carried out as follows: 10 cm^2 pieces of interfacing are sterilised and placed in petri dishes. An aqueous solution of filter-sterilised prepolymer (10% w/v) is mixed with a spore suspension in the ratio of 5:1 (w/w). Under sterile conditions this mixed suspension is added dropwise (0.6 g suspension / 10 cm^2) to the interfacing. The impregnated pieces of fabric are allowed to dry at room temperature in the dark for 8–10 h before being exposed to daylight for 6 h, followed by a 10 min irradiation with a high-pressure mercury lamp (150 W; 366 nm). The impregnated pieces of fabric are now ready for use.

Immobilisation in Microcapsules by Polyelectrolyte Complex Formation

Cellulose sulphate (degree of substitution 0.2–0.6) is the polyanion and polydimethyldiallylammonium chloride (PDMDAAC) serves as the polycation. 100 ml of a 4% (w/v) solution of cellulose sulphate is prepared and autoclaved. 10 ml of a spore suspension is added after cooling. The suspension is then added dropwise to 300 ml of an autoclaved 2% solution of PDMDAAC under stirring as described for Ca alginate immobilisation, and the capsules formed are left to stand for 30 min. The capsules are removed by filtration, washed three times with sterile 0.9% NaCl and are then ready for use.

All solutions and equipment used for immobilisation must be sterile!

Use of β-Cyclodextrin as a Solvent for Steroids

In order to reduce or avoid the use of toxic solvents, cyclodextrins (CD) can be used as solubilisers. Cyclodextrins are cyclic derivatives of glucose, in which the glucose molecules are β-linked via 1-4 bonds. They are produced by bacteria (e.g. *Bacillus macerans*) as a result of starch breakdown. Three types, α- β- and γ-CD are found in which 6, 7 or 8 glucose molecules are linked. α-CD is the most water-soluble of the three, but for reasons of economy β-CD will be used. CDs have the ability to take up substances, even lipophilic ones, into the spaces resulting from their spatial arrangement and transport them into aqueous systems. They are usually resistant to hydrolysis.

The aim of using β-CD in this experiment is to achieve a better turnover of the steroids and to reduce the amount of methanol required. The following must be carried out:

1   *Checking the breakdown/non-breakdown by the strain used*: addition of spores to the medium containing β-CD as the sole carbon source. The growth rate and the amount of glucose produced can be determined.

2   *Determination of the β-CD content of the medium*:
   (i)   *Solutions*: β-CD (5×10^{-4} M = 56.75 mg/100 ml H_2O); phenolphthalein solution (3.75×10^{-3} M = 0.119 43 mg/100 ml 96% ethanol) = solution 1. Solution 1 is diluted 1:10 with H_2O to give solution II (must be made up freshly!); Na_2CO_3 solution (4×10^{-2} M = 4.24 g/1000 ml H_2O).
   (ii) *Measurement*: Calibration values of β-CD (e.g. 0.5, 1.0, 1.5, 2.0, 2.5, 5.0 ml) or samples of the culture medium are treated as follows:
      (a)   0.5–5.0 ml sample/standard
      (b)   2.0 ml phenolphthalein solution II
      (c)   2.5 ml Na_2CO_3 solution
      (d) H_2O ad 25.0 ml

      OD_{550} is measured against H_2O within 10 min using a cuvette with a 1 cm light path. The temperature should be 25 °C ± 2 °C.

3   *Comparison of steroid turnover* with and without β-CD.

Interpretation

The parameters to be considered in the interpretation of the experiment are listed in Table 3.12. In the individual variations the reduction in substrate or the increase in product concentration is to be determined and entered in the table. The turnover (%) can be calculated from the amount of product formed and the amount of substrate used. If necessary the reduction in the concentration of cyclodextrin should also be noted.

The dry weight of the cells, as a measure of growth, can be determined as follows: for non-immobilised cells the contents of three flasks are filtered through dried and weighed filters. The filter cake is washed three times with distilled water. For immobilised cells the dry weight of three aliquots of 50 Ca alginate beads or microcapsules or three pieces of impregnated interfacing should be taken. The filters are dried at 90 °C and then weighed. The difference in weight between incubated and

Table 3.12 Hydroxylation of steroids by free or immobilised fungal cells: *S*, steroid substrate concentration; *P*, product concentration; *U*, turnover rate; *DW*, dry weight of the biomass; *Glc*, glucose concentration; *β-CD*, *β*-cyclodextrin concentration; $Y_{x/s}$, yield coefficient.

t (h)	*S* (g/l)	*P* (g/l)	*U* (%)	*DW* (g/l)	*Glc* (g/l)	*β-CD* (g/l)	$Y_{x/s}$
0							
24							
48							
72							

non-incubated samples of the immobiliser allows the amount of growth to be calculated. For biotransformations with growing cells the amount of glucose used can be determined with the aid of a commercially available test. For growth on glucose or cyclodextrin the yield coefficient should be given.

Further reading

CHINOIN, Budapest, Hungary and AVEBE, Veendam, Netherlands (1987) *Molecular encapsulation by cyclodextrins*. Manual.

GRÖGER, D. and JOHNE, S. (1982) *Mikrobielle Gewinnung von Arzneimitteln-Pharmazeutische Mikrobiologie* (Berlin: Akademie Verlag).

KIESLICH, K. (1984) Introduction. In: H.-J. Rehm, and G. Reed (eds) *Biotechnology – a comprehensive treatise in 8 volumes*, Vol. 6a, *Biotransformations*, pp. 1–4 (Weinheim: Verlag Chemie).

KIESLICH, K. (1984) Biotransformationen. In: P. Präve, U. Faust, W. Sittig and D.A. Sukatsch (eds) *Handbuch der Biotechnologie* (München: R. Oldenbourg Verlag).

KIESLICH, K. (1991) Biotransformations of industrial use. *Acta Biotechnol.* **11**, 559–70.

SZEJTLI, J. (1981) The metabolism, toxicity and biological effects of cyclodextrins. In: D. Duchenne (ed.) *Cyclodextrins and their industrial use*, pp. 173–210 (Paris: Editions de Santé).

VIKMON, M. (1981) Rapid and simple spectrophotometric method for detection of micro-amounts of cyclodextrins. *Proc. 1st Int. Symp. on Cyclodextrins, Budapest*, pp. 69–74.

WOERDENBAG, H.J., PRAS, N., FRIJLINK, H.W., LERK, C.F. and MALINGRE, T.M. (1990) Cyclodextrin facilitated bioconversion of 17*β*-estradiol by a phenoloxidase from *Mucuna pruriens* cell cultures. *Phytochem.* **29**, 1551–4.

YAMADA, H. and SHIMIZU, S. (1988) Mikrobielle und enzymatische Verfahren zur Produktion biologisch und chemisch wertvoller Verbindungen. *Angew. Chemie* **100**, 640–61.

4

Isolation Techniques

4.1 SEPARATION OF SOLID PARTICLES

H. Schütte

The example chosen is harvesting cells of *Saccharomyces cerevisiae* by cross-flow filtration.

Principle and Aim of the Experiment

Cross-flow filtration is an alternative method to the classical solid/liquid separation techniques such as sedimentation, centrifugation, separation or filtration. Micro-filtration membranes with a pore diameter between 0.2 μm and 0.45 μm, either as planar membranes in a flat holder or folded or spiral membranes in the appropriate holders, are most often used and in some instances tubular membranes in the form of hollow fibres or capillaries. The method described here using micropore membranes is a means of obtaining cells from fermenter cultures. The concentration of biomass is in principle a cyclic process in which the biomass is concentrated to the extent required for the next step in the process.

Description of the Experiment

In this experiment the harvesting of cells from a 10 l fermenter will be carried out. The filter module used can accommodate filters with a surface area of 700 cm^2. The cell suspension is pumped at high speed through the filter module so that it crosses the membrane, which is orientated at right angles to the direction of flow. The material retained on the filter is returned to the reserve tank and the filtrate is collected. The *transmembrane pressure (TMP)*, the force responsible for cross-flow filtration, is regulated by adjusting the pressures at the entry and exit of the cell suspension to and from the filter module and in the filtrate collection vessel. During the filtration process data for flux (filtration capacity in $l\ h^{-1}\ m^{-2}$), cell concentration, TMP (calculated from the individual pressures) and the velocity of the pump are taken against time, and the cross-flow filtration is stopped as soon as the required amount of biomass has been obtained. A schematic representation of a cross-flow-filtration set-up is shown in Fig. 4.1.

Principle

In classical filtration (cake filtration, static filtration, dead-end filtration) the total volume of cell suspension to be filtered is pressed through the filter at a defined pressure. A filter 'cake' is formed which in itself acts as a filter, with the result that the velocity of filtration falls continuously when all other parameters remain unchanged. The aim of cross-flow filtration is, by means of appropriate flow technology, e.g. parallel flow across the membrane placed at right angles to the direction of filtration, to prevent the build-up of a filter 'cake'. Fig. 4.2 illustrates the principal difference between dead-end filtration and cross-flow filtration (CFF). Under ideal conditions the resistance to filtration in CFF is dependent only on the filter material used. The cross-flow principle can be achieved by two means:

1   the cell suspension is pumped through the membrane at high speed
2   by continual movement of the membrane and/or holder (shear filter).

Ultrafiltration or micropore membranes are used as the filter material. The pressure across the membrane is defined as transmembrane pressure and is calculated as follows:

$$TPM = \frac{pressure_{entry\ point} + pressure_{exit\ point}}{2} - filtrate\ pressure$$

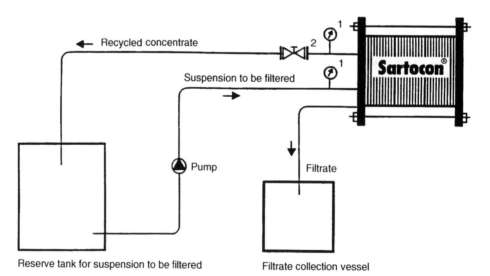

Figure 4.1   Schematic representation of a cross-flow-filtration system: 1, membrane manometer; 2, membrane valve.

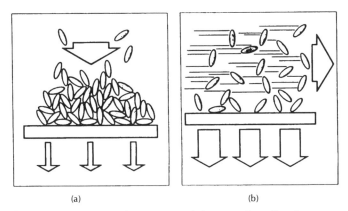

(a)   (b)

Figure 4.2 Schematic comparison of (a) static and (b) cross-flow filtration.

Microorganism, Media, Materials and Apparatus

- *Microorganism.* Baker's yeast (*Saccharomyces cerevisiae*) from wholesaler. The experiment described here can be applied to the harvesting of cells from a fermentation of 10 or 20 l.

- *Media.* When starting from a 10 l fermentation no medium is required. If bulk yeast is used, the cells should be resuspended in 50 mM potassium phosphate buffer, pH 7.0 and the pH re-adjusted with 0.1 M KOH.

- *Materials.* Filters which can be used are a 700 cm^2 Omega cassette, pore size 0.16 or 0.3 μm (Filtron) or comparable filters from Millipore or Sartorius.

- *Apparatus.* For example Filtron Minisette and pump, stopwatch and measuring cylinder.

Procedure

Unless a 10 l fermentation has been carried out, prepare a 3% suspension of yeast in buffer and adjust the pH to 7.0. The container with the yeast suspension is placed on ice and stirred throughout the experiment. Before switching on the pump, the connections to and from the membrane are fully opened and the connection to the filtrate collection vessel closed. The yeast suspension is pumped, as shown in Fig. 4.1, through the membrane at a speed of 150–200 l/h. After a few minutes the connection to the filtrate collection vessel is opened slowly. The transmembrane pressure is set at 0.8–1.5 bar and the flux determined every 250 ml up to 1 l and then every 500 ml. At the same time the OD$_{600}$ and the dry weight of the yeast is determined in both the starting suspension and in every litre of filtrate obtained. The volume of the cell suspension should be reduced to 500 ml, if possible. The connections to and from the membrane should be opened fully, and that to the filtrate collection vessel should be closed and the concentrated cell mass forced out of the membrane with buffer. OD$_{600}$ and the dry weight of the final suspension is determined. The membrane is regenerated by washing with buffer and then cleaned according to the manufacturer's instructions.

Conclusions

The values for flux, OD$_{600}$ and dry weight are plotted against the volume of the filtrate. The effects of altering the membrane pressure or the speed of the pump

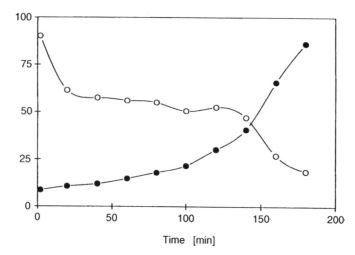

Figure 4.3 Concentration of an *E. coli* fermentation by cross-flow filtration (200 l, 0.2 μm polypropylene cassette filter): o—o l/(h m²); •—• *DW*, g/l.

should be discussed and the rate of concentration achieved in the final sample is to be calculated. The typical course of the flux curve should be discussed and suggestions made as to how the average flux can be increased when the experiment is repeated. Fig. 4.3 illustrates the concentration of an *E. coli* fermentation culture by cross-flow filtration.

Further reading

GRABOSCH, M. (1987) Cross-Flow-Filtration in der Biotechnik. *BioEngineering* **1**, 62–8.
KRONER, K.H. (1986) Zum Einsatz der Querstromfiltration bei der Aufarbeitung von Enzymen. *BTF – Biotech-Forum* **3**, 22–31.
SCHULZ, G. (1986) Mikrofiltration – ein Verfahren in der Biotechnologie. *BTF – Biotech-Forum* **3**, 136–40.

4.2 CELL DISRUPTION

H. Schütte

4.2.1 *Determination of the 'Disruption Constant' with a Bead Mill*

Principle and Aim of the Experiment

With a bead mill it is possible to break open all types of microorganisms in a continuous process. If no data are available for the microorganism to be broken open, it is useful to start with the following conditions:

- glass beads (lead-free) 0.5 nm diam.
- volume to which the disruption vessel should be filled with glass beads: 85%
- rotation velocity of the grinding discs: 10 m/s

- flow rate for yeast: 10 × volume of the grinding vessel / h
- flow rate for bacteria: 5 × volume of the grinding vessel / h
- cell concentration: unimportant.

If the results of one run are unsatisfactory (i.e. <60% of the cells are broken) the first three variables can be optimised. For this purpose one determines the 'disruption constant' (k) in batch cultures. During the experiment the release of the desired product is measured with respect to time. If the breakage vessel has no openings to allow samples to be taken, the experiment can be performed using an externally fitted beaker. An increase in the value of k means that the correct choice for the value of a particular variable has been made for the organism in question. In batch culture, data are also obtained which allow the flow rate to be optimised. If the time required for total breakage is 3 min, the time for continuous flow should be approximately 10 min because there is no 'plug flow' in a bead mill. The working values obtained for a specific microorganism are valid only for the grinder and container used; i.e. another model or even volume requires re-calibration.

Procedure

Since the determination of k in batch culture requires an opening in the container for taking samples and the entire experiment would only take 5 min, for the purpose of this book the experiment will be altered. Diagrams of the bead mill and some of its component parts are shown in Fig. 4.4. The experimental apparatus to be used is shown in Fig. 4.5.

The bead mill is filled, avoiding bubbles, with some of a known volume of stirred yeast suspension. The bead mill is switched on and the cell suspension is pumped into the bead mill at a constant rate. The expelled suspension is returned to the starting suspension. From the beginning of the experiment, small samples are taken at defined time intervals and tested for the protein or enzyme in question. The experiment should be carried out for 1 h, and the final sample should correspond to the maximum amount of desired product released. The maximum amount of protein released is an important factor in calculating the breakage rate constant and should therefore be determined and confirmed in a separate analytical experiment.

Principle of Cell Disruption in a Bead Mill

In the grinding vessel there is a high-speed stirrer whose circular velocity ranges from 5 to 15 m s^{-1}. 80–85% of the volume of the grinding vessel is taken up with lead-free glass beads 0.5 μm in diam. The force of movement is transferred from the grinding disc to the glass beads mainly in the form of adhesion forces combined with repulsion forces which are in turn dependent on the arrangement and type of grinding discs. The acceleration of the grinding blades in a radial direction causes layers of movement. The differential velocities between the layers of glass beads result in high shearing forces related to the absolute velocity and size of the moving beads which, together with collisions, are responsible for breaking the walls of the microorganisms. The differential velocities are also determined by the distances from the fixed parts of the apparatus.

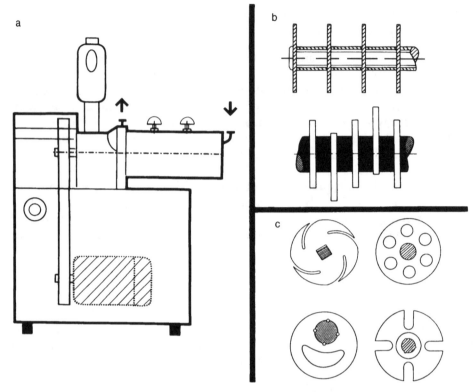

Figure 4.4 (a) Diagram of a bead mill; (b) grinding mechanism with centric and acentric positioning of the grinding discs; (c) examples of grinding discs used for cell disruption.

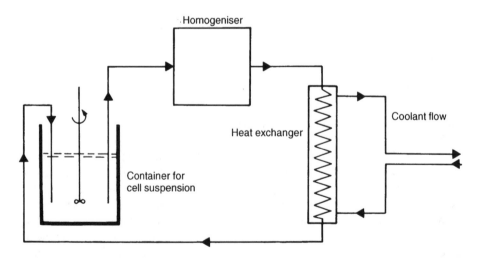

Figure 4.5 Experimental set-up for cell disruption in a continuous system.

Theoretical Considerations on the Coefficient of Disintegration

Several groups have shown by experiment that cell breakage is a reaction following first-order kinetics, i.e. the rate at which the product is produced is at any time proportional to the amount of product available.

$$\frac{dP}{dt} = -kP$$

where k is the disruption constant and P is the product concentration at time t. Integration and correction with respect to the amount of product at a specific time gives the following:

$$\ln\left(\frac{P_m}{P_m - P}\right) = kt$$

where P_m is the disruption amount of product released.

This formula applies when the k value is determined by direct sampling from the grinding vessel. In the example described above, where the sample is removed from without the grinding vessel, the formula is altered as follows:

$$\ln\left(\frac{P_m}{P_m - P}\right) = kN$$

where N is the number of times the suspension has passed through the mill and is calculated as follows:

$$N = \frac{\text{Flow volume}}{\text{Initial volume}}\, t$$

Microorganism, Equipment and Solutions

- *Microorganism*: baker's yeast or another microorganism grown in the laboratory
- *Bead mill*: 0.5 mm glass beads (lead-free), funnel, stopwatch, 5 l beaker, magnetic stirring bar, magnetic stirrer, 100 ml measuring cylinder, compressed air and water supply for cooling
- *Analysis*: protein analysis (see pp. 10–11)
- *Buffer*: 100 mM potassium phosphate buffer, pH 7.5.

Outline of the Experiment

1.6 kg of yeast is resuspended in 100 mM potassium phosphate buffer, pH 7.5 and brought to a final volume of 4 l (40% w/v). The pH of the suspension is determined and, if necessary, re-adjusted to 7.5 by the addition of 100 mM potassium hydroxide. Place 850 ml dry glass beads into the mill vessel (volume: 1 l) and then fill with the cell suspension. The remaining yeast suspension is stirred continuously on ice and connected to the input and output valves of the mill vessel. The mill vessel should be run at a speed of 2000 rpm and a flow volume of 10 l per hour. Throughout the experiment the grinding vessel should be cooled. Samples are taken

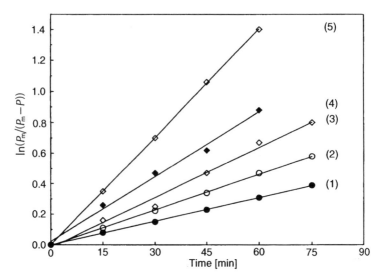

Figure 4.6 Determination of the disruption constant in batch method of a 40% yeast suspension using five different grinding discs: (1) 12 eccentric; (2) 12 eccentric; (3) 6 conical; (4) 7 conical; (5) 8 conical.

at time 0, 1, 2, 5, 7.5, 10, 15, 20, 30, 45 and 60 min. Throughout the experiment the temperature, pH and conductivity of the yeast suspension are monitored. At the end of the experiment the remaining cell suspension is forced out of the mill vessel by running water through it and subsequently emptying it. The glass beads are dried at 100 °C.

Interpretation and Conclusions

After the maximum amount of protein released and the amounts released at time t have been determined, the values for

$$\ln\left(\frac{P_m}{P_m - P}\right)$$

are plotted against time as shown in Fig. 4.6 for five different bead mills. The slope of each straight line is the breakage constant.

In another graph the values for temperature, pH and conductivity are plotted against time.

The diagrams should be interpreted and conclusions drawn.

N.B. The experiment can be carried out using a pressure homogeniser.

Further reading

LIMON-LASON, J., HOARE, M., OSBORN, C.B., DOYLE, D.J. and DUNNILL, P. (1979) Reactor properties of a high-speed bead mill for microbial cell rupture. *Biotechnol. Bioeng.* **21**, 745–74.

MARFFY, F. and KULA, M.-R. (1974) Enzyme yields from cells of brewer's yeast disrupted by treatment in a horizontal disintegrator. *Biotechnol. Bioeng.* **16**, 623–34.

MELENDRES, A.V., HONDA, H., SHIRAGAMI, N. and UNNO, H. (1991) A kinetic analysis of cell disruption by bead mill. *Bioseparation* **2**, 231–6.

SCHÜTTE, H. and KULA, M.-R. (1986) Einsatz von Rührwerkskugelmühlen und Hochdruckhomogenisatoren für den technischen Aufschluß von Mikroorganismen. *BTF-Biotech-Forum* **3**, 68–80.

SCHÜTTE, H. and KULA, M.-R. (1990) Pilot- and process-scale techniques for cell disruption. *Biotech. Appl. Biochem.* **12**, 599–620.

SCHÜTTE, H. and KULA, M.-R. (1990) Bead mill disruption. In: J.A. Asenjo (ed.) *Separation processes in biotechnology*, pp. 107–41 (New York: Marcel Dekker).

SCHÜTTE, H. and KULA, M.-R. (1993) Cell disruption and isolation of non-secreted products. In: H.-J. Rehm, G. Reed and G. Stephanopoulos (eds) *Biotechnology*, Vol. 3: *Bioprocessing*, pp. 505–26 (Weinheim: VCH).

4.2.2 *Optimisation of the Disruption of* Saccharomyces cerevisiae *using a High-pressure Homogeniser*

Theory and Aim of the Experiment

In order to obtain microbial intracellular products, e.g. enzymes, it is necessary to disrupt the microorganisms chemically (Naglak *et al.*, 1990), biologically (Andrews and Asenjo, 1987) or mechanically (Engler, 1990; Hetherington *et al.*, 1971; Schütte and Kula, 1986; Schütte and Kula, 1993). Practical experience has shown that high-pressure homogenisation and grinding in a bead mill are universally applicable methods. The disruption process must be carefully monitored because unopened cells represent a loss of the desired intracellular product. When carrying out mechanical cell disruption it is very important to ensure that any increase in temperature during homogenisation is compensated for as soon as possible, usually by circulating cold water through the cooling jacket. Another important point is to monitor the pH during cell breakage and to adjust it to the value of the starting suspension. Depending on the microorganism, 1–3 passages at maximum pressure (70–120 MPa) results in >95% breakage. After breakage the cell extract is subjected to a solid/liquid separation step, then either a precipitation or an extraction step.

In the experiment described below, a 40% (w/v) suspension of *Saccharomyces cerevisiae* will be subjected to high-pressure homogenisation resulting in the release of protein, including the enzyme fumarase. The aim of the experiment is twofold: to determine the optimal pressure for homogenisation and, by means of several cycles of homogenisation, to find out the maximum amounts of total protein and fumarase released. Since the homogenisation pressure has an effect on the output of the homogenisers, the flow rate should be determined for each different pressure. The intracellular substances released during cell disruption can alter the pH and the conductivity of the cell homogenate; it should be noted whether or not these changes influence the efficiency of cell disruption.

Principle

A high-pressure homogeniser consists of a pressure pump and a homogeniser. Depending on the model the required pressure can be set either manually or automatically. The cell suspension is brought to the pressure set by means of the pressure pump. The valve is opened by about 0.075 mm when the pressure rises above that set; this causes a sudden and very rapid (280 ms) drop in the pressure of the cell suspension, which rushes (approx. 350 m/s) through the open valve (Fig. 4.7). The resulting turbulence and shearing forces, and the direct collision with the pressure ring, positioned at right angles to the direction of flow, result in disruption of the cell walls.

Microorganism, Media and Equipment

- *Microorganism*: baker's yeast (*Saccharomyces cerevisiae*)
- *High-pressure homogeniser*: Gaulin homogeniser with a maximum pressure of 100 MPa and a flow volume of 40 l/h or similar. The homogeniser must be fitted with a cell disruption valve and a cooling jacket
- *Analysis*:

    Fumarase: see precipitation of proteins, p. 145

    Protein: see analytical cell disruption, p. 10–11
- *Additional equipment*: spectrophotometer, conductivity meter, pH meter, stopwatch, thermometer, measuring cylinder (1 l), 10 l beaker, magnetic stirring bar and a fine household sieve
- *Buffer*: 100 mM potassium phosphate buffer, pH 7.5, 100 mM potassium hydroxide.

Procedure

4 kg of yeast is resuspended in 100 mM potassium phosphate buffer, pH 7.5, and brought up to a final volume of 10 l (40%w/v). The pH of the suspension is measured and, if necessary, re-adjusted to 7.5 by the addition of potassium hydroxide.

In the meantime, using water at room temperature, the flow rates at different pressures and the accompanying increase in the temperature of the water (without cooling) are determined for working pressures between 0 and 1000 bar (Fig. 4.8).

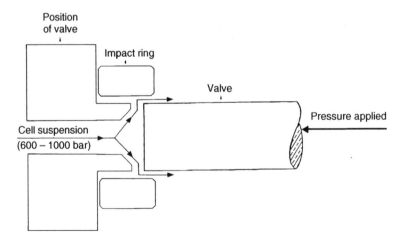

Figure 4.7 Cross-section through a Gaulin-CD homogenisation valve during cell breakage.

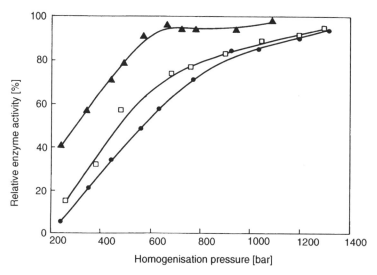

Figure 4.8 Disruption of three different microorganisms as a function of homogenisation pressure. ● *Saccharomyces cerevisiae,* □ *Bacillus cereus,* ▲ *E. coli*

The yeast suspension is poured through the sieve into the reservoir of the homogeniser and the cells are disrupted at pressures between 1 and 100 MPa in steps of 10 MPa. In contrast to the 'trial run' using water, this time the cooling system is switched on. The pressure is set and 15 s later a sample of 100 ml is taken. The following data should be collected:

1 before the experiment:
   (i)   temperature of the cell suspension
   (ii) pH of the cell suspension at 20 °C
   (iii) conductivity of the cell suspension at 20 °C
2 during the experiment:
   (i)   the temperature of the homogenate at the outlet valve
   (ii) the flow rate
3 at the end of the experiment:
   (i)   fumarase activity
   (ii) protein content
   (iii) pH of the cell suspension at 20 °C
   (iv) conductivity of the cell suspension at 20 °C.

To determine the maximum cell disruption, three runs are performed at 1000 bar.

Interpretation and Conclusions

The following curves should be drawn:

• water temperature v. pressure
• water temperature per 10 MPa

- temperature of the yeast suspension v. pressure
- flow rate of water v. pressure
- reduction in flow rate per 10 MPa
- flow rate of the yeast suspension v. pressure
- fumarase activity v. pressure
- protein content v. pressure
- rate of disruption (fumarase activity and protein content) v. pressure
- changes in pH v. pressure
- changes in conductivity v. pressure
- changes in pH and conductivity v. rate of disruption.

References

ANDREWS, B.A. and ASENJO, J.A. (1987) Enzymatic lysis and disruption of microbial cells. *TIBTECH* **5**, 273–7.

ENGLER, C.R. (1990) Cell disruption by homogenizer. In: J.A. Asenjo (ed.) *Separation processes in biotechnology*, pp. 95–105 (New York: Marcel Dekker).

HETHERINGTON, P.J., FOLLOWS, M., DUNNILL, P. and LILLY, M.D. (1971) Release of protein from baker's yeast by disruption in an industrial homogenizer. *Trans. Inst. Chem. Eng.* **49**, 142–8.

NAGLAK, T.J., HETTWER, D.J. and WANG, H.Y. (1990) Chemical permeabilization of cells for intracellular product release. In: J.A. Asenjo (ed.) *Separation processes in biotechnology*, pp. 177–205 (New York: Marcel Dekker).

SCHÜTTE, H. and KULA, M.-R. (1986) Einsatz von Rührwerkskugelmühlen und Hochdruckhomogenisatoren für den technischen Aufschluß von Mikroorganismen. *BTF-Biotech-Forum*, **3**, 68–80.

SCHÜTTE, H. and KULA, M.-R. (1993) Cell disruption and isolation of non-secreted products. In: H.-J. Rehm and G. Reed (eds) *Biotechnology*, Vol. 3: G. Stephanopoulos (ed.) *Bioprocessing*, pp. 505–26 (Weinheim: Verlag Chemie).

4.2.3 *Small-scale Disruption of Microorganisms*

Theory and Aim of the Experiment

Disruption of small quantities of cells in order to measure enzyme activities is one of the routine tasks in a microbiology laboratory. Cell disruption is a prerequisite, for example, for monitoring fermentations, for the preparation of growth curves, for optimisation of media and for screening enzymes. Small-scale cell disruption can be performed easily using a mixing mill. In contrast to other analytical cell disruption methods (see Table 4.1) a mixing mill can, depending on the size of the disruption vessel and the holder used, cope with between 2 and 16 samples simultaneously.

The length of time required for disruption varies from 5–15 min depending on the organism. The increase in temperature during the disruption process is <5 °C, meaning that cooling is unnecessary. Treatment in a mixing mill is a reliable, time-saving method of cell disruption which requires no extra cooling system.

Table 4.1 Small-scale disruption methods.

Apparatus	Principle	Method
French press	high-pressure homogenisation	Pressure: 1400 bar
X-press	low temperature, high pressure	Pressure: 2000 bar Temp.: $-35\,°C$
Sonicator	sonication	Frequency: 20 kHz
Parr bottle	Nitrogen decompression	Pressure: 155 bar

Description of the Experiment

A suspension of *Saccharomyces cerevisiae* will be disrupted in a mixing mill, using disruption vessels with a volume of 12.5 ml and the rate of disruption determined by measuring the activity of the cytoplasmic enzyme, glucose-6-phosphate dehydrogenase, and the amount of protein released over a period of time. The optimal conditions for disruption will be determined by varying the size of the glass beads used, the disruption time and the relative amounts of cell suspension and glass beads.

Principle

The lead-free glass beads in the disruption vessel are subjected to oscillation frequencies between 150 and 1800 min^{-1} (max.). The acceleration of the glass beads causes a build-up of layers with differing flow rates. Depending on the absolute velocity and size of the glass beads, shear forces are created and these, together with collision events, are responsible for the disruption of the cells.

Microorganism, Media and Equipment

- *Microorganism*: commercially available baker's yeast (*Saccharomyces cerevisiae*) is used.
- *Buffer*: the yeast cells are resuspended in 100 mM potassium phosphate, pH 7.5. The pH of the cell suspension should be adjusted to 7.5 with 0.1 M potassium hydroxide and re-adjusted again after disruption.
- *Equipment*: lead-free glass beads in various sizes.
- *Glucose-6-phosphate-dehydrogenase test*:

  0.1 M Triethanolamine buffer, pH 7.6 (adjust with

1 N NaOH)	2600 μl
+ MgCl$_2$ (2 g MgCl$_2$ × 6 H$_2$O/100 ml)	200 μl
+ NADP$^+$ (10 mg/ml)	100 μl
+ H$_2$O	3000 μl

  The enzyme sample is added and, after mixing, the activity is measured at 340 nm (extinction coefficient: 6.3 cm^2/μmol) and 25 °C.
- *Protein determination according to Bradford*: Bovine serum albumin is used as the standard. Method: see p. 10–11.

Procedure

Prepare a 40% (w/v) suspension of yeast cells in 0.1 M potassium phosphate buffer, pH 7.5, and re-adjust the pH to 7.5 with 0.1 M KOH. Place 10.0 g glass beads (diam. 0.5 mm) and 5.0 g yeast suspension in one disruption vessel and 10.0 g glass beads (diam. 1.0 mm) and 5.0 g yeast suspension in a second disruption vessel. The vessels are sealed and placed in the holders before switching on the oscillation mill at maximum frequency (1800 min^{-1}) and setting the timer on 'hold'. A sample (200 μl) is taken from the starting suspension, and samples of the same volume are withdrawn after 1, 3, 6, 10, 15 and 30 min via the sampling valve. To avoid changes in the volume, 200 μl of the initial suspension are fed into the disruption vessel at each of the sampling times. The samples taken up to 10 min after the start of the experiment are used to illustrate the increase in the extent to which the cells are disrupted, whereas the values of the 15 and 30 min samples should correspond to the maximum degree of disruption and they provide the '100%' value, with which the values of the other samples may be compared. This experiment can be carried out using glass beads of diameters different from those given, different amounts of glass beads and yeast suspension. The oscillation frequency of the mill and the concentration of the cell suspension can be varied and the whole experiment can be performed with a microorganism other than yeast. This experiment allows the optimal conditions of cell disruption to be determined for analytical purposes (Table 4.2).

Interpretation and Conclusions

The following graphical representations of the data are to be drawn:

1   protein concentration (mg/ml) as a function of time
2   enzyme content (U/ml) as a function of time

Table 4.2   Disruption of a 20% (w/v) yeast suspension using glass beads with diameters of either 0.75 mm or 1.0 mm.

Time	*Saccharomyces cerevisiae* activity released (%)				Protein (%)			
	a	*b*	*c*	*d*	*a*	*b*	*c*	*d*
0	2	2	2	2	2	2	2	2
5	24	87	67	41	25	100	70	45
10	45	100	85	58	47		82	60
15	70		94	67	66		90	65
30	100		100	93	100		100	81

Frequency of oscillation 1800 min^{-1}.
[a] 10.6 g glass beads + 6.5 ml cell suspension.
[b] 10.0 g glass beads + 5.0 ml cell suspension.
[c] 7.5 g glass beads + 5.0 ml cell suspension.
[d] 5.0 g glass beads + 5.0 ml cell suspension.

3 degree of disruption (%) as measured by protein concentration and enzyme activity as a function of time (the values obtained with the samples taken at 30 min are set as 100%)

4 change in temperature as a function of time

5 degree of disruption (%) as a function of:

  (i)   the diameter of the glass beads used

  (ii) amount of glass beads/amount of cell suspension

  (iii) concentration of the cell suspension.

When plotting the graphs in (5) the values obtained at 2, 5 and 10 min should be used.

Bibliography

BRADFORD, M.M. (1976) A rapid and sensitive method for the quantitation of microgram quantities of protein utilizing the principle of protein-dye binding *Anal. Biochem.* **72**, 248–54.

HUMMEL, W. and KULA, M.-R. (1989) Simple method for small-scale disruption of bacteria and yeasts. *J. Microb. Meth.* **9**, 201–9.

SCHÜTTE, H. and KULA, M.-R. (1988) Analytical disruption of microorganisms in a mixer mill. *Enzym. Microb. Technol.* **10**, 552–8.

4.3 PRODUCT ENRICHMENT

H. Schütte

4.3.1 *Precipitation of Proteins from a Crude Yeast Extract by Ammonium Sulphate*

Theory and Aim

Protein precipitation is one of the first steps towards product recovery. The purpose of precipitation with a neutral salt such as ammonium sulphate or sodium sulphate is either concentration or fractionation of the desired product. When fractionation is performed there is often a 2–5 fold enrichment of the product because different proteins are precipitated at different concentrations of ammonium sulphate. The aim of the following experiment is to construct ammonium sulphate precipitation curves for the enzymes fumarase, glucose-6-phosphate dehydrogenase and the proteins of a 40 % (w/v) crude extract of *Saccharomyces cerevisiae*. Using the information in the curves obtained it should be discussed whether or not a separation of the selected proteins by ammonium sulphate is possible.

Description of the Method

Finely ground ammonium sulphate is added slowly and under continual stirring to a crude yeast extract to 40% (w/v) saturation. Stirring is continued for a given time before the suspension is centrifuged. Enzyme activity and protein concentration of the supernatant are determined. Total enzyme activity and protein concentration of the crude extract are compared with the values obtained after ammonium sulphate precipitation.

Principle

The protein precipitation obtained at high concentrations of ammonium sulphate is the result of a decrease in the hydration of proteins in favour of that of the ions of the neutral salt. The hydration layer surrounding the surface of the protein is transferred to the ions of the salt at a rate depending on the charge of the protein. The concentration required to precipitate a specific protein is dependent on the salt used and is also a function of pH and temperature. The reproducibility of a protein precipitation is also dependent on the rate at which the salt is added [2–10 g/(1 min)], the length of time the suspension is stirred and whether the salt is added as a finely ground powder or as a saturated solution of ammonium sulphate. The use of a saturated solution of ammonium sulphate is gentler on the protein but is only to be recommended for small volumes since the volume of the sample doubles when 50% saturation is reached.

Microorganism, Media and Equipment

1 *Microorganism.* Commercially available baker's yeast is used for the preparation of the crude yeast extract.

2 *Preparation of the crude yeast extract.* Baker's yeast is resuspended in 100 mM potassium phosphate buffer, pH 7.5, at a final concentration of 40% (w/v). The pH is re-adjusted to 7.5 with 0.1 M potassium hydroxide after suspension of the yeast and again following the cell disruption in either a high-pressure homogeniser or in a bead mill. The homogenate obtained is centrifuged for 30 min at 25 000 *g*, and the supernatant, which is cloudy, is decanted and re-centrifuged under the same conditions. The supernatant from the second centrifugation is the crude extract to be used for the experiment described below.

3 *Determination of fumarase activity.* 25 ml 1 M potassium phosphate buffer is added to 3.352 g/l malic acid and the pH adjusted to 7.3 with 1 M KOH. The volume is made up to 500 ml with deionised water. This is the stock solution which can be stored frozen in small aliquots. Enzyme assay: stock solution 1000 μl; enzyme solution (may be diluted if necessary) 50 μl. The test is performed at 30 °C and the change in extinction at 250 nm is measured (coefficient of extinction: 1.45 cm^2/μmol).

4 *Measurement of glucose-6-phosphate dehydrogenase.* See p. 143.

5 *Determination of protein concentration with Coomassie-Blue G250: Bradford method.* Bovine serum albumin is used as the standard protein (see p. 10–11).

6 *Other requirements.* Finely ground ammonium sulphate; refrigerated centrifuge; 200 ml beakers; magnetic stirrer; stopwatch and pH meter.

7 % saturation of ammonium sulphate (A); amount in g of ammonium sulphate to be added to 1 l protein solution (B):

A:	10	15	20	25	30	35	40	45	50	55
B:	56	84	114	144	176	209	243	277	313	351

A:	60	65	70	75	80	85	90	95	100
B:	390	430	472	516	561	610	662	713	767

Procedure

An ammonium sulphate precipitation with the following degrees of saturation (room temperature) is to be performed: 30, 35, 40, 45, 55, 60 and 70%.

For each degree of saturation, place a 200 ml beaker containing 100 ml crude extract and a stirring magnet on a magnetic stirrer. Using the information given under step 7 above, calculate the amounts of ammonium sulphate required to bring each 100 ml aliquot to the given degree of saturation. Under constant stirring add the finely ground ammonium sulphate at the rate of 5 g/(l min) to the crude extract. After the addition of the total amount of the salt the pH is measured and adjusted to 7.5 with 1 M KOH. Stirring is continued for 1 h to attain the correct equilibrium. The final volume is measured before centrifuging for 30 min at 15 000 *g*. The supernatant is decanted and the volume measured before determining the fumarase and glucose-6-phosphate dehydrogenase activities and the protein concentration. For comparative purposes these measurements must also be made for the crude extract before the ammonium sulphate precipitation (100% values). The result of this experiment is shown in Fig. 4.9. The pellet is taken up in 50 ml 0.1 M potassium phosphate buffer, pH 7.5, stirred for 30 min and then centrifuged. Each pellet should contain the remaining amount of protein required to make up 100% as determined after the ammonium sulphate precipitation. The values for the enzyme activities and protein concentrations depend on the volume and must be multiplied by the volume to calculate the total amounts. Using a larger volume of crude extract, e.g. 500 ml, an attempt should be made to separate the fumarase and glucose-6-phosphate dehydrogenase activities by precipitation at 45% saturation (see Fig. 4.9).

Interpretation

A graph similar to that shown in Fig. 4.9 should be plotted and a diagram drawn to illustrate the discovery of the enzyme and/or protein in the pellet. The results

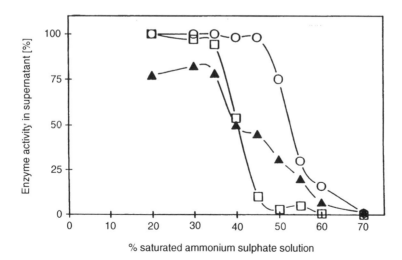

Figure 4.9 Activities of the two intracellular enzymes fumarase and glucose-6-phosphate dehydrogenase and protein concentration at different degrees of ammonium sulphate saturation of a 40% crude extract of *Saccharomyces cerevisiae*: o—o glucose-6-phosphate dehydrogenase; □—□ fumarase; ▲—▲ protein.

147

obtained with the 500 ml aliquot should be compared with those obtained from the small-scale experiment and the findings discussed. Furthermore the yield obtained after an ammonium sulphate precipitation and the purification factor for each of the enzymes should be noted.

4.3.2 Desalting and Conditioning a Salt-containing Protein Solution by Cross-flow Filter Dialysis

Principle and Aim

Following precipitation with a neutral salt or column chromatography, proteins exist in solutions with a high salt concentration. For later purification steps the protein solution has to be conditioned, i.e. the amount of salt reduced and possibly placed in another buffer system. On a small scale this can be done in dialysis tubing. For larger volumes gel filtration or filter dialysis allow efficient desalting and change of buffer.

Description of the Technique

The salt concentration of a protein solution is reduced by continuous cross-flow filter dialysis in a recycle batch system. For filter dialysis either a plate module or a hollow fibre module is used. During filter dialysis the volume of the protein solution which is being desalted must be kept constant by the addition of water or buffer, thus ensuring an effective removal of the salt. The experimental set-up for filter dialysis is shown in Fig. 4.10. Generally five times the volume of the original protein solution is sufficient

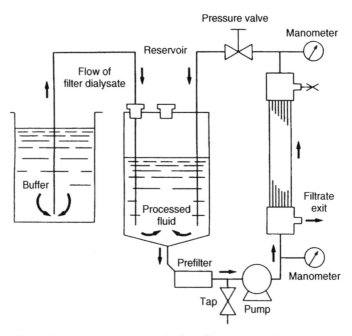

Figure 4.10 Dialysis filtration in an Amicon hollow-fibre system (alternative option to plate module).

to reduce the salt concentration to the desired level. The decrease in salt concentration can be followed by continually monitoring the conductivity of the solution.

Principle

In filter dialysis, membranes are used which are impermeable to protein molecules but do not hinder the passage of smaller molecules. Membranes are particularly useful for the exchange of salts or sugars or for the removal of low-molecular-weight substances; the ionic strength and the pH of the solution can be altered at the same time. The rate of de-salting is the same as the filtration rate because the salt transport is a convection process. This rate is only a function of the ultrafiltration rate and is independent of the concentration of the substances being separated. The filter dialysis performed here is done in a cross-flow module. The aim of cross-flow filtration is, by means of appropriate flow mechanics, e.g. parallel flow across the membrane combined with an effective intermittent gradient acting at right angles to the direction of filtration, to prevent the build-up of a secondary membrane of protein.

Media and Equipment

- *Protein solution*. It is possible to use a particle-free crude extract, an eluate from a chromatography column, the resuspended pellet from an ammonium sulphate precipitation, or the filter dialysate from a cell homogenate.
- *Buffer*. 50 mM potassium phosphate, pH 7.5, a similar buffer of low molarity, or deionised water should be used for filtration dialysis.
- *Conductivity measurement*. The conductivity meter should be coupled to a measuring device allowing the conductivity to be measured continually.

Procedure

Ammonium sulphate or sodium chloride is added to 1000 ml protein solution until the conductivity of the solution has a value of about 100 mS/cm. The cross-flow-filter dialysis is performed in a Filtron-Minisette system (Omega filter cassette with a filter area of 700 cm^2; cut-off 10 kDa). The pump should deliver about 40 l/h, and the pressure across the membrane should lie between 0.8 and 1.0 bar. The device for measuring the conductivity is placed in the high-salt protein solution so that the reduction in conductivity (ion concentration) can be monitored continuously throughout the filtration. The high-salt protein solution is maintained at a constant volume by the addition of fresh buffer to compensate for the loss generated by the production of filtrate. Throughout the filter dialysis the flux should be determined as a function of time; this is done by measuring the time required to collect 200 ml filtrate. The filter dialysis is continued for as long as is necessary to reach the desired conductivity. The volume of buffer required is at least six times that of the protein solution. The graph in Fig. 4.11 illustrates the course of a typical cross-flow-filter dialysis.

Interpretation and Discussion

A curve as shown in Fig. 4.11 should be drawn. In a second diagram the flux [l/(h m^2)] should be plotted against time to determine whether or not the flux alters. The surface

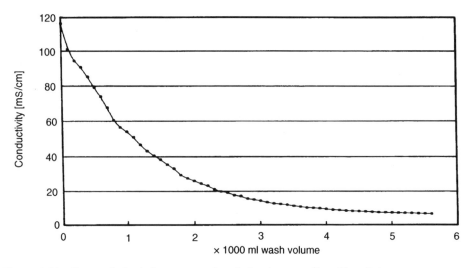

Figure 4.11   Removal of salt from a protein solution by cross-flow-filter dialysis.

area of the membrane relative to the stage of the experiment should be taken into account. The experiment can be carried out at various membrane pressures and/or pump speeds. Discussion points are the number of 'wash volumes' (relative to the volume of the protein solution) required, the influence of membrane pressure / pump speed and the time involved as compared with other methods, e.g. gel filtration or dialysis.

4.4   PRODUCT RECOVERY

H. Schütte

4.4.1   *Separation of the Enzymes Glucose-6-phosphate Dehydrogenase and Fumarase by Ion Exchange Chromatography*

Theory and Aims

Ion exchange chromatography is frequently chosen as the first fine purification step in the separation and purification of biomolecules. The reasons for this are easy handling, good separation, high capacity, speed (making it useful for large sample volumes) and, in comparison with other chromatographic methods, its low cost. By selecting the appropriate pH value the ion exchanger can be used either to concentrate a particular protein or for separating a mixture of proteins. Whether an anion or cation exchanger is chosen as the stationary phase depends on the isoelectric point of the desired protein and the pH value. Most proteins have a negative charge at a physiological pH so that a DEAE (diethylaminoethyl-), QAE (quaternary aminoethyl-) or Q (quaternary amine) exchanger can be used. Basic proteins can be separated on cation exchangers with CM (carboxymethyl-) or SP (sulphopropyl-) as the ligand.

Description of the Experiment

Following cell disruption, centrifugation, ammonium sulphate precipitation and filter dialysis, the enzymes glucose-6-phosphate dehydrogenase (G-6-P DH) and fumarase from *Saccharomyces cerevisiae* will be bound to a strongly basic anion exchanger and then eluted by means of a linear salt gradient, i.e. they will be purified separately. It will then be tested if they can be separated by a stepwise elution. The data obtained provide information for planning a scaled-up experiment which will be carried out at the same linear flow rate. After the column chromatography a 'cleaning' *in situ* will be performed.

Principle

The ion exchanger consists of a matrix, e.g. cellulose or cross-linked agarose, to which charged ligands and exchangeable ions of opposite charge are chemically bound. A matrix with cationic ligands (e.g. DEAE) contains exchangeable anions and is therefore termed an 'anion exchanger'. Proteins must have a surplus charge of negative ions to bind to an 'anion exchanger'; this means that the pH of the protein solution must be considerably greater than that of the isoelectric point. The negative charge arises from the carboxyl groups of glutamic acid and aspartic acid and to the carboxyl group at the C-terminal end of the protein, whereas the positive charge is contributed to by the basic amino acids arginine, lysine and histidine as well as the amino terminus of the polypeptide. The charged state of the acidic and basic amino acids depends on the pH value: a low pH (an excess of H^+ ions) prevents the dissociation of the acidic groups; at a high pH the basic groups are uncharged. Protein bound to an anion exchanger can be eluted by an increasing salt gradient or by a descending pH gradient since the charge decreases towards the isoelectric point. In general it can be said the greater the charge of the protein to be isolated the higher the ionic strength of the elution buffer should be.

Microorganism, Equipment and Media

- *Microorganism*: commercially available baker's yeast (*Saccharomyces cerevisiae*)
- *Anion exchanger*: Q-sepharose FF from Pharmacia or another strong basic anion exchanger from another manufacturer
- *Equipment*: chromatography column (5/30) fitted with an adaptor; peristaltic pump (adjustable to 500 or 1000 ml/h); UV monitor (280 nm filter); fraction collector with space for 100 tubes connected to a recording device; pH meter and conductivity meter
- *Enzyme tests*:
  - G-6-PDH – see analytical cell disruption, p. 143
  - fumarase – see protein precipitation, p. 146
  - protein – see analytical cell disruption, p. 145–148

Procedure

A chromatography column with a diameter of 5 cm and a length of 30 cm is packed with Q-sepharose FF according to the manufacturer's instructions so that the height of the material is about 15 cm. After removing the air from beneath the grid and in the tubing the adaptor is placed, avoiding air bubbles, on top of the column and the column is equilibrated. This is done by adjusting the pH of the column to exactly 7.5 by passage of 500 mM potassium phosphate buffer, pH 7.5, at a rate of 50 ml/h, then the column is washed with equilibration buffer (50 mM potassium phosphate buffer, pH 7.5) to adjust the conductivity to about 8 mS/cm (linear velocity = ml h^{-1} cm^{-2}). 100 ml protein solution obtained as described in Sections 4.2 and 4.3 is now applied to the column at a linear flow rate of 15 cm/h. It is essential that the sample has the same pH and conductivity as the ion exchanger. After the sample has been loaded, the column is washed with one 'column volume' of equilibration buffer or with as much as is needed until the UV monitor shows that the base line has been reached. All the buffer used for washing the column is collected and tested for G-6-P DH and fumarase activity. The bound enzymes are eluted with a linear gradient (see Fig. 4.12).

The salt gradient is set up as follows. Three 'column volumes' each of equilibration buffer and 'limit' buffer (50 mM potassium phosphate buffer, pH 7.5 + 0.5 M NaCl) are placed in separate 1 l beakers and the beaker containing the equilibration buffer is placed on a magnetic stirrer. The other beaker is placed on a 'lab jack' at a higher level, and a 'bridge' of tubing allows the 'limit' buffer to flow into the equilibration buffer. The column is now washed with the salt gradient by pumping buffer from the beaker containing equilibration buffer into the column at a rate of 25 cm/h.

20 ml fractions are collected in the fraction collector and the eluted protein is followed on the recorder. Every fifth fraction is tested for enzyme activity; in the region where the desired enzyme elutes, every fraction is tested. The fractions

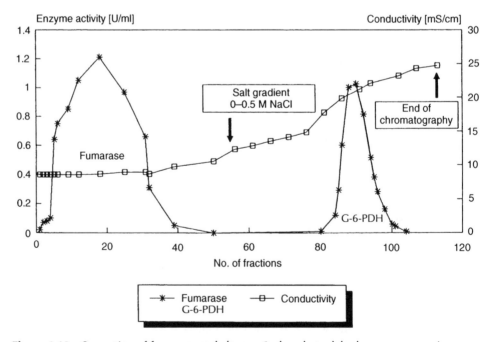

Figure 4.12 Separation of fumarase and glucose-6-phosphate dehydrogenase on an ion exchanger.

showing activity are combined and the activity and protein content determined. The ion exchanger can be regenerated by washing the column with one 'column volume' of 50 mM potassium phosphate buffer, pH 7.5 + 1 M NaCl and then washing with equilibration buffer to prepare the column for the next run. In contrast to the first run, this time the two enzymes will be separated by appropriate step elution. If a larger column (diam. 10 cm) is available a 'scaled-up' experiment using the same linear flow rate as in the previous experiment can be carried out. When an experiment is 'scaled up' the parameters column material, length of the column, linear flow rate, buffer composition, ionic strength and pH remain constant. The remaining parameters – e.g. sample volume and concentration, volumes of washing and elution buffers or the gradient – are increased proportional to the volume of the column. After several chromatography runs the column should be 'cleaned up'. The ion exchanger is cleaned with sodium hydroxide as described by the manufacturer, washed to neutrality with deionised water and then equilibrated as described at the beginning of the experiment.

Interpretation and Conclusions

Curves should be drawn for each chromatography run in which the number of fractions is drawn on the X-axis and the values for activity, OD_{280}, pH and conductivity are drawn on the Y-axis. Total enzyme activity before and after the column run, yield, specific activity and the purification factor should be determined. The efficiency of separation, the advantages and disadvantages of gradient and step elution, and the result of the 'scale-up' experiment are to be discussed.

Further reading

Ion exchange chromatography – principles and methods (Pharmacia).
ERIKSSON, K.-O. (1989) In: J.-C. Janson and L. Ryden (eds) *Protein purification – principles, high resolution methods and applications* (Weinheim: VCH).
SHUEY, C.D. (1990) Ion-exchange processes. In: J.A. Asenjo (ed.) *Separation processes in biotechnology*, pp. 263–86 (New York: Marcel Dekker).

4.4.2   *Separation of the Enzymes Glucose-6-phosphate Dehydrogenase and Fumarase by Hydrophobic Interaction Chromatography (HIC)*

Theory and Aim of the Experiment

After a salt precipitation or ion exchange chromatography where the protein is eluted by a salt gradient, the protein solution has a high salt concentration. To avoid 'conditioning' without a purification effect, e.g. desalting by gel filtration or filter dialysis, it is possible to perform *hydrophobic interaction chromatography (HIC)*. HIC utilises the different hydrophobicities of proteins and peptides. This type of chromatography relies on interactions between the sample and the hydrophobic groups on the separation matrix. The hydrophobicity of the sample is increased by the addition of a salt, e.g. ammonium sulphate, or the sample is already in an ammonium sulphate solution following one of the purification steps described above. Under these conditions the proteins are adsorbed onto the hydrophobic carrier and can be eluted with a decreasing salt gradient relative to their hydrophobicity at low ionic strength. HIC can be used to concentrate protein solutions or for fractionating proteins as the first in a series of chromatography purification steps, after other chromatographic procedures or as one of the later steps of a protein purification.

Description of the Experiment

Following cell disruption, centrifugation and ammonium sulphate precipitation the enzymes glucose-6-phosphate dehydrogenase and fumarase from *Saccharomyces cerevisiae* are separated by HIC on phenyl-sepharose CL-4B. The enzymes are eluted in a decreasing linear ammonium sulphate gradient. Once the system has been optimised it is attempted to elute the enzymes with a step rather than a gradient elution. At the end of the chromatography the matrix will be regenerated and equilibrated again. The experimental set-up is the same as for ion-exchange chromatography.

Principle

There are hydrophobic regions on the surface of biological molecules which become masked by hydrophilic or ionic groups when the molecules are placed in a low-ionic-strength milieu. If the ionic strength is increased, e.g. by the addition of ammonium sulphate, the hydrophilic interaction of water and the biomolecule is reduced. The hydrophobic regions on the surface of the biomolecule can interact with the hydrophobic ligands on the matrix in the presence of high salt. Therefore proteins in a buffer of high ionic strength are adsorbed onto the hydrophobic ligands; a reduction in the ionic strength of an eluant causes a stronger interaction between the protein and the water molecules and as a consequence the protein is eluted from the hydrophobic matrix. This method causes only weak conformational changes in the protein and thus allows a high yield of the desired product. The salt concentration chosen depends on the hydrophobicity of the protein being purified; the concentration of ammonium sulphate used is in the range 1.0–1.7 M. The strength of the hydrophobic interactions is affected by the following series of ions:

Increasing hydrophobicity

$\longleftarrow$

Anions: PO_4^{3-}, SO_4^{2-}, CH_3COO^-, Cl^-, Br^-, NO_3^-, ClO_4^-, I^-, SCN^-

Cations: NH_4^+, Rb^+, K^+, Na^+, Cs^+, Li^+, Mg^{2+}, Ca^{2+}, Ba^{2+}

$\longrightarrow$

Increasing chaotropy

Parameters other than the amount and type of salt influence hydrophobic interactions. As well as a reduction in ionic strength, lowering the temperature, increasing the pH, the addition of organic solvents, chaotropic salts, detergents or ethylene glycol can weaken hydrophobic interactions.

Microorganism, Solutions and Apparatus

- *Microorganism*: commercially available baker's yeast
- *HIC chromatography material*: phenyl-sepharose CL-4B or octyl-sepharose CL-4B from Pharmacia or similar HIC material from another manufacturer

- *Apparatus*: chromatography column (2.6/30) with adaptor; peristaltic pump adjustable to 500 ml/h; UV monitor (280 nm filter); fraction collector for 100 tubes with recording device; pH meter and conductivity meter
- Finely ground ammonium sulphate
- *Enzyme tests*:
  - G-6-PDH – see p. 143
  - fumarase – see p. 146
  - protein – see p. 10–11.

Procedure

A column of the size given above is packed to a height of about 25 cm with phenyl-sepharose CL-4B according to the manufacturer's instructions. The column is then equilibrated with 50 mM potassium phosphate buffer, pH 7.5 and 1.0 M ammonium sulphate at a flow rate of 300 ml/h. When the pH and the conductivity values of the eluate and the equilibration buffer are identical, the column may be regarded as equilibrated. If the *S. cerevisiae* enzyme sample has not been precipitated with ammonium sulphate this salt should be added, with stirring, to a final concentration of 1.0 M. In either case the pH and the conductivity of the sample should be adjusted to those adopted for equilibration of the column. 60 ml of enzyme suspension (pH 7.5, conductivity 125 mS/cm) is pumped on to the column at a flow rate of 100 ml/h. The column is then washed with two column volumes (OD_{280} drops to virtually zero) of buffer at a flow rate of 300 ml/h. The wash buffer is collected in a beaker and this is assayed for the activity of the enzyme being purified. The column is eluted with a decreasing ammonium sulphate gradient. This is done by mixing five volumes of 'limit' buffer (50 mM potassium phosphate, pH 7.5) and five volumes of equilibration buffer as described for ion exchange chromatography, except that in this experiment the solution is pumped out of the beaker containing the buffer with the higher salt concentration at a flow rate of 300 ml/h. The fraction collector is also switched on and fractions of 20 ml are collected by time. The elution can be followed on the recording device. The conductivity and the enzyme activity is measured in every fifth fraction to determine when the enzyme is eluted. The conductivity and enzyme activity is then measured in every fraction around the elution point. The 'active' fractions are pooled and the total activity and the specific activity are measured. The separation of the enzymes can be optimised by increasing the salt concentration of the equilibration buffer and the sample or by altering the gradient or by performing a stepwise elution. The column matrix can be regenerated by washing with deionised water and then re-equilibrated with equilibration buffer. After several runs or when discoloration is observed at the point of loading, the column should be regenerated according to the manufacturer's instructions.

Interpretation and Discussion

Graphs should be plotted for OD_{280}, enzyme activity and the conductivity of the column eluate. Total activity before and after the column run, the respective yield, specific activity, purification factor and the conductivity range in which the enzymes are eluted are all to be calculated. The separation effect should be discussed and suggestions made for improving the separation.

Further reading

Octyl- and phenyl-sepharose CL-4B. Information brochure (Pharmacia).
ERIKSSON, K.-O. (1989) Hydrophobic interaction chromatography. In: J.-C. Janson and L. Ryden (eds) *Protein purification principles, high resolution methods and applications*, pp. 207–27 (Weinheim: Verlag Chemie).

4.5 FERMENTATION AND PROCESSING

4.5.1 *Production of Citric Acid by* Aspergillus niger: *an Example of the Large-scale Microbial Production of a Bulk Product and its Purification*

R. Buchholz

Principle and Aim of the Experiment

Bioproduction can be regarded as a combination of fermentation and processing techniques. The two methods should be considered as a single production unit because of the competition afforded by chemical synthesis. The choice of production method is dictated, among other things, by the effect of energy metabolism on the product formation. According to Gaden there are three types of microbial production. The purification method depends on the concentration of the end product, the nature and concentration of any side products and on the lability of the end product.

Production and processing methods will be demonstrated taking the production of citric acid as an example. Insight will be gained into the management and processing of a compound produced on a large scale. On the basis of the information obtained from the fermenter run the process will be classified according to Gaden. At the end of the fermentation the initial industrial production steps and waste products thereby accumulated will be demonstrated and parameters affecting the yield will be discussed.

Experimental Procedure

Citric acid obtained from the fermentation of sucrose by *Aspergillus niger* is removed from the culture by precipitation. A 10 l batch fermentation is carried out. Growth of the strain involves culturing on plates, harvesting the conidia, and setting up a shaken flask culture for inoculation of the fermentation vessel. The fermentation is monitored by removing samples for the determination of the amounts of substrate, biomass and product. At the end of the fermentation the product is precipitated out of the solution.

Media

- Standard agar for yeast and fungi:
glucose	50.0 g/l
KH_2PO_4	2.0 g/l

MgSO$_4$ × 7 H$_2$O	1.0 g/l
peptone	8.0 g/l
yeast extract	2.0 g/l
agar	20.0 g/l
dH$_2$O ad 1000 ml	

- Preculture medium:

sucrose	40.0 g/l
KH$_2$PO$_4$	0.2 g/l
NH$_4$NO$_3$	0.2 g/l
MgSO$_4$ × 7 H$_2$O	0.3 g/l
FeSO$_4$ × 7 H$_2$O	0.1 mg/l
ZnSO$_4$	0.25 mg/l
methanol	3.0% (v/v)
dH$_2$O ad 1000 ml	

- Fermentation medium:

refined sugar	150.0 g/l
NH$_4$NO$_3$	2.0 g/l
MgSO$_4$ × 7 H$_2$O	0.25 g/l
KH$_2$PO$_4$	0.5 g/l
CuSO$_4$	5 ppm
K$_4$Fe[CN]$_6$	15 ppm
polypropylene glycol	0.025 ml/l

Media Preparation

All media should be sterilised at 121 °C for 20 min. For the preculture medium the ferrous sulphate solution is filter-sterilised and then added to the sterile medium in the flask. The N$_2$ source is added after 18 h of growth; this causes the development of mycelia in the form of small pellets consisting of short, fat hyphae. For the preculture medium the sugar solution should be sterilised separately to avoid the formation of Maillard products which have a detrimental effect on yields.

For production purposes the sugar solution and the potassium hexacyanoferrate are sterilised together in the fermenter. After cooling to 30 °C the salt solution (autoclaved separately) is added and the pH of the medium is adjusted to 2.0–2.5 with 10% H$_2$SO$_4$.

Cultivation

Spores taken from an agar slant are streaked onto a standard agar plate. The plate is incubated at 30 °C, and after 2 days black conidia start to form on the surface of the white mycelia. Three days later the black conidia are washed off with a 1% solution of NaCl. The conidial suspension is collected in a 50 ml Erlenmeyer flask and the spore titre is determined.

Four 1 l Erlenmeyer flasks, each containing 100 ml preculture medium, are inoculated with 3×10^7 spores per flask and incubated at 30 °C on a reciprocal shaker (110 rpm) for 18 h. The nitrogen source is now added and the incubation continued for a further 54–78 h.

The fermenter is inoculated with the combined precultures. The fermentation is run for 200–300 h at 30 °C and an initial stirring speed of 800 rpm and an aeration rate of 1 vvm. The pH should be monitored and the oxygen saturation should not fall below 25%. Samples are withdrawn every 12 h and used to determine the dry weight, sugar content and the amount of product produced. The culture should also be examined microscopically.

Laboratory Measurements

- *Dry weight determination.* Dry weight is determined either by differential weighing or with a commercially available measuring device.
- *Substrate determination.* This is performed using the test kit for sucrose determination available from Boehringer Mannheim.
- *Product determination.* Citric acid production is followed with the help of the enzymatic test kit available from Merck (BDH).
- *Processing.* Mycelia are harvested from the medium by means of a separator, rotating drum filter or by centrifugation. 100 ml of the filtrate obtained is filtered a second time. Citric acid is precipitated from this solution by the addition of 180–250 g/l $Ca(OH)_2$ at 70 °C. The pH of the solution is adjusted to 7.2 ± 0.2. Chalk milk is added to the solution under continuous stirring at a ratio of 0.4–0.8 relative to citrate. The precipitated tri-calcium citrate tetrahydrate is collected by filtration and washed several times with distilled water. Tri-calcium citrate tetrahydrate is brought into solution by the addition of 80% H_2SO_4 at 60 °C. The sulphuric acid should be in a 0.2 vol% excess. At this stage the precipitation of calcium sulphate can be observed. The concentration of citric acid in the solution is determined.

Microorganism, Media and Equipment

- *Microorganism: Aspergillus niger* ATCC 11414
- *Laboratory analysis*: in addition to the test kits described above the following is required: filter apparatus, drying oven, photometer, cuvettes, microscope, Thoma cell counting chamber, pipettes
- *Fermentation*: 10 l fermentation vessel (glass) fitted with stirrer and controls for regulating and measuring temperature, aeration, stirring speed, pH and dissolved oxygen
- *Processing*: separator, rotating drum filter or centrifuge; thermostatically controlled magnetic stirrer, thermometer, $Ca(OH)_2$, conc. H_2SO_4.

Calculation and Interpretation

The information collected during the experiment should be presented in the form of graphs. In order to be able to regulate the process it is absolutely essential that the experimental values and calculated data are noted as soon as they have been determined. From the curves obtained the type of production according to Gaden can be identified. On the basis of the results obtained, alternative methods of processing should be suggested and discussed. The loss sustained in the purification step carried out should be determined. The discussion should take into account at which stage the loss occurred, factors affecting the processing and what possibilities there are for improvement.

Further reading

CRUEGER, W. and CRUEGER, A. (1990) In: T.D. Brock (ed.) *Biotechnology: a textbook of industrial microbiology* (Sunderland, MA: Sinauer Associates).

GADEN, E.L. (1955) Fermentation kinetics and productivity. *Chem. Ind.*, 154–9.

GADEN, E.L. (1955) Fermentation process kinetics. *J. Biochem. Microbiol. Techn. Eng.* **1**, 413–29.

KUBICEK, C.P. and RÖHR, M. (1986) Citric acid fermentation. *Crit. Rev. Biotechnol.* **3/4**, 331–73.

REHM, H.-J. (1982) *Industrielle Mikrobiologie* (Berlin: Springer Verlag).

REHM, H.-J. and REED, G. (1983) In: H. Dellweg (ed.) *Biotechnology*, Vol. 3 (Weinheim: VCH).

SODECK, G., MODL, J. KOMINEK J. and SALZBRUNN, W. (1981) Production of citric acid according to the submerged fermentation process. *Proc. Biochem.*, **16**(6), 9–11.

4.5.2 *Microbial Production and Processing of the Polysaccharide Schizophyllan*

P. Götz

Principle and Aim of the Experiment

Polysaccharides of microbial origin with constant product characteristics can be produced on a technical scale for various purposes. Polysaccharides are required in many fields, including the food industry where they are used as thickening and stabilising agents, for the pharmaceutical industry which uses dextran as a blood plasma substitute, the building industry where the polymers improve the rheological properties of plaster, emulsion paints and adhesives and in the oil-refining industry. The biopolymer schizophyllan is used in Japan for manufacturing an airtight film for packaging of fruit and vegetables.

For the food industry, not only the functional properties of the biopolymer are important but also issues such as the total elimination of cell material from the biopolymer before it is used. Furthermore the production strain must be non-pathogenic and incapable of toxin production. For these reasons production strains should be GRAS organisms, i.e. generally regarded as safe.

The strain chosen for this experiment, *Schizophyllum commune*, grows as a parasite on the branches of trees or as a saprophyte on fallen wood. This fungus, considered to be one of the white rot fungi, possesses the metabolic ability to break down lignin. When the strain grows on wood, the less well-favoured substrate cellulose remains unchanged. The basidiomycete *Schizophyllum commune* is suitable for biopolymer production for the following reasons:

- it is non-pathogenic for man

- it is easily cultured under laboratory conditions

- since the polysaccharide is excreted into the medium there is no need for hyphal disruption.

Schizophyllan is a homopolysaccharide consisting of β-1,3-linked glucose molecules with a monomer joined in the 1,6 position in every third link of the polymer chain. It is water-soluble, uncharged and stable over a wide pH range. Under normal growth conditions *Schizophyllum commune* does not produce schizophyllan. The fungus excretes the polysaccharide only when growing in submerged culture on a suitable carbon source, e.g. glucose. Once the primary carbon source has been exhausted the polysaccharide is metabolised again.

Description of the Experiment

Principle

Downstream processing of microbially produced polysaccharides takes advantage of the particular flow characteristics of the fermentation liquor – high viscosity at a low product concentration and non-Newtonian flow behaviour. The product can be separated from the biomass by filtration. Cross-flow filtration can be used because the viscosity of the polysaccharide solution decreases with increasing shear forces.

The method of choice for the further processing of polysaccharides is their precipitation from an aqueous solution by an organic solvent. The addition of the organic solvent reduces the dielectric constant, which in turn reduces the solubility. The solvent chosen should be miscible with water at all concentrations and as inexpensive as possible. For safety reasons closed systems must be used, thus increasing costs. These can be offset by using a solvent which can be recovered by distillation. Taking all these points into account, precipitation is frequently done with ethanol or isopropanol. Important parameters for the precipitation are the alcohol concentration, temperature and the nature and concentration of the salt used in addition to the alcohol. The addition of a salt means that less alcohol has to be added, thereby reducing production costs. The best yields for xanthene are obtained with 65% ethanol and 1% KCl at 15 °C.

Organism and Media

MICROORGANISM

Schizophyllum commune CBS 266.60 – classification of the fungus: Class, Basidiomycetes; Subclass, Homobasidiomycetes; Order, Agaricales; Race, Schizophyllum; Species, commune.

MEDIA

- Solid medium for cultivation (malt extract agar, pH 5.3):
  malt extract   30.0 g/l
  agar–agar   20.0 g/l
  dissolved in dH_2O.
- Liquid medium for polysaccharide production, pH 5.3:
  glucose   30.0 g/l
  KH_2PO_4 1.0 g/l
  $MgSO_4 \times 7\,H_2O$ 0.5 g/l
  yeast extract   3.0 g/l
  dissolved in dH_2O.

The glucose-salts and yeast extract solutions should be autoclaved separately.

Cultivation

Cultivation of the fungus for the submerged culture is carried out on malt extract agar at 30 °C. The plate is fully grown after 7 days. Two pieces (each about 1 cm^2) of the mycelial lawn serve as the inoculum for 200 ml medium in a 1 l Erlenmeyer flask. The pieces of mycelia are shredded by homogenisation of the flask contents with a double-bladed metal stirrer at a speed of 1200 rpm for 1 min. Incubation is carried out with shaking (90 rpm) at 26 °C.

Procedure

Fermentation liquor from cultures of the basidiomycete *Schizophyllum commune* are processed to obtain schizophyllan. Product concentration and the productivity of the process are determined. Changes in the biomass and substrate concentrations should be monitored throughout the fermentation.

   The procedure takes several days; therefore care must be taken that a sufficient number of samples can be processed. An autoclavable fermenter with a capacity of at least 10 l is required. It is somewhat easier to perform the experiment using flask cultures; for each time point, two flasks are required (duplicate samples). Each of the required number of flasks, containing 200 ml medium, are inoculated with 5 ml homogenised suspension from the preculture vessel. In order to obtain curves which can be interpreted, samples should be taken at regular intervals. This can be done by staggering the times of inoculating the flasks. For the experiment illustrated in Fig. 4.13, 12 of the 24 flasks required were inoculated 12 hours later than the others, thus allowing simultaneous processing of samples with two different incubation times. The schedule below permits four flasks per day to be processed without any effect of the staggered inoculation being observed.

1   The total content of the flask is passed through a dried, weighed filter to obtain the cells.

2   The filtrate is stored at 4 °C for further treatment.

3   The cells are washed with distilled water to remove any traces of medium before being placed in a drying oven for 12 h at 105 °C. The cells are then placed in an exsiccator and allowed to cool for 2–3 h before being weighed.

4   An aliquot of the filtrate is mixed with three times its volume of 98% ethanol and left to stand at 4 °C for 2 h to precipitate the polysaccharide.

5   The suspension from the previous step is transferred to dried and weighed centrifuge tubes and centrifuged at 25 000 *g* for 30 min. After decanting the supernatant the tubes are dried at 100 °C for 12 h, placed in an exsiccator to cool for 2–3 h and then weighed.

6   The glucose concentration in an aliquot from step 2 is determined enzymatically using a commercially available test kit.

7   The biomass and product concentrations can be calculated from the differences in weight measured for the filter and the tubes in steps 3 and 5, respectively, and the respective sample volumes.

Interpretation and Conclusions

The experimental data obtained should be noted carefully and a graph drawn as illustrated in Fig. 4.13. If several groups perform the experiment then all data should

161

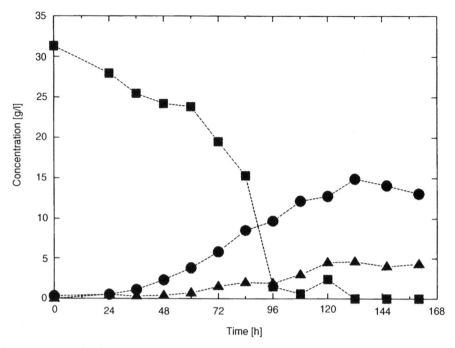

Figure 4.13 Curve for the production of schizophyllan from *Schizophyllum commune*: ● biomass; ■ substrate; ▲ product.

be collected so that they can be interpreted by all groups. The productivity is calculated from the amount of product obtained in a given time.

A detailed mathematical description is obtained by combining a logistic growth model and a product kinetic according to Luedeking and Piret. By an approximation of the model parameter it is possible to determine to what extent schizophyllan production is coupled to the growth of the fungus. The equation for the logistic growth of biomass in a batch process is:

$$\frac{dX}{dt} = r_X = \mu X \left(1 - \frac{X}{X_{max}} \right) \tag{4.1}$$

where r_X is the rate at which biomass is produced, μ is the specific growth rate and X_{max} is the maximum amount of biomass which can be obtained.

The growth-dependent amount of product r_P is obtained with the parameter α, and the growth-independent amount of product is obtained with the parameter β:

$$\frac{dP}{dt} = r_P = \alpha r_X + \beta X \tag{4.2}$$

The change in the substrate concentration with respect to time is dependent on two terms, one representing the amount of substrate used for growth and the other the amount required for product formation:

$$\frac{dS}{dt} = -r_S = -\frac{1}{Y_{X/S}} r_X - \frac{1}{Y_{P/S}} r_P \tag{4.3}$$

It has been shown that this type of model for describing different processes is well suited for the microbial production of exopolysaccharides. Calculation of the values of the model parameters μ, X_{max}, α, β, $Y_{X/S}$ and $Y_{P/S}$ based on the figures obtained for biomass-, substrate- and product concentrations from a batch culture are possible with the help of a numerical method (solution of DGL first order, combined with a Simplex method). A graphical presentation of such data with similar model considerations is shown in Weiss and Ollis (1980) and Klimek and Ollis (1980).

References

KLIMEK, J. and OLLIS D.F. (1980) Extracellular microbial polysaccharides: kinetics of *Pseudomonas* sp., *Azotobacter vinelandii* and *Aureobasidium pullulans* batch fermentations. *Biotechnol. Bioeng.* **22**, 2321–42.

WEISS, R.M. and OLLIS, D.F. (1980) Extracellular microbial polysaccharides, 1: Substrate, biomass and product kinetic equations for batch xanthan gum fermentation. *Biotechnol. Bioeng.* **22**, 859–73.

4.5.3 *Purification of Malate Dehydrogenase by Affinity Chromatography from a Crude Extract of Saccharomyces cerevisiae*

H. Schütte

Principle and Aim of the Experiment

Specific interactions between biologically active substances, such as enzymes and their substrates, cofactors or inhibitors, between antigen and antibody, glycoproteins and lectins, etc. provide the basis for affinity chromatography. In order to carry out affinity chromatography it is essential to bind the appropriate affinity ligand via a spacer to the gel matrix. Passage of a heterologous sample through such a column results in the 'selective' binding of molecules which have an affinity for the matrix-bound ligand. After removal of all unbound molecules the bound molecules can be eluted specifically by an exchange reaction or via a soluble ligand or unspecifically by altering the conformation of the molecule by alteration of pH or ionic strength. If elution is not possible by any of these methods a more drastic approach can be taken, e.g. elution by denaturing compounds such as thiocyanate or similar chaotropic substances. On account of the high selectivity of the ligands, affinity columns, in contrast to other chromatography columns, can be used only for limited purposes. The advantage of affinity chromatography lies in its specificity, because high purification factors are therefore easily achieved. Affinity chromatography can be carried out at virtually any stage of a protein purification; however, the high costs of the column matrix and the elution compounds must be taken into consideration.

Description of the Experiment

Malate dehydrogenase (MDH) will be obtained from a partially purified (cell disruption, centrifugation and ion exchange chromatography) protein extract of *Saccharomyces cerevisiae* by affinity chromatography on Blue Sepharose CL-6B (Pharmacia). MDH will be eluted from the column by the relatively inexpensive method of increasing the ionic strength of the elution buffer. Initially the elution will be performed with a linear gradient and, once the elution conditions have been determined, in a second attempt with a specifically chosen step gradient. After the affinity column has been used, the matrix will be regenerated and re-equilibrated according to the manufacturer's instructions.

Theory of the Technique

A number of proteins are capable of being adsorbed onto Blue Dextran. Dyes such as Cibacron Blue F3GA or Procion Red HE3B are competitive inhibitors of oxido-reductases which require NAD^+ or $NADP^+$ as a cofactor. Three-dimensional models of the dye Cibacron Blue F3GA and NAD^+ show the similarity of the substances, in particular the ADP and ribose parts of the molecule but not in the nicotinamide moiety. Cibacron Blue F3GA can act as an analogue of AMP, ADP, ATP, IMP and GTP. On account of their relaxed specificity, immobilised dyes such as Cibacron Blue F3GA are termed 'group specific' materials. Fig. 4.14 illustrates the use of an affinity column with Blue Sepharose CL-6B for the fractionation of a crude extract of *Saccharomyces cerevisiae*. As shown in the illustration, in addition to a specific elution it is also possible, by altering the pH value and/or the ionic strength, to elute the cofactor to which the enzyme has the highest affinity. Even when only a single cofactor is used for the elution, the elution of any enzyme can be controlled by means of the concentration of the cofactor. The purification factors and yields for the named examples are given in Table 4.3.

Microorganism, Media and Apparatus

- *Microorganism*: commercially available baker's yeast (*Saccharomyces cerevisiae*)
- *Affinity chromatography matrix*: Blue Sepharose CL-6B (Pharmacia) or similar material from another supplier

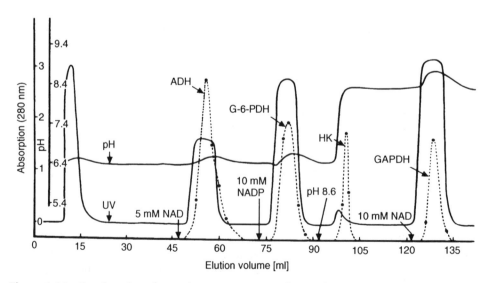

Figure 4.14 Fractionation of a crude yeast extract on Blue Sepharose CL-6B.

Table 4.3 Data for purification of several enzymes from a crude extract of *Saccharomyces cerevisiae* on Blue Sepharose CL-6B (see Fig. 4.14).

Enzyme	Purification factor	Yield (%)
Alcohol dehydrogenase (ADH)	31 ×	55
Glucose-6-phosphate dehydrogenase G-6-PDH	4460 ×	60
Hexokinase (HK)	10 ×	25
Glyceraldehyde-3-phosphate dehydrogenase (GAPDH)	155 ×	55

- *Chromatography equipment*: chromatography column (16/30) and adaptor; peristaltic pump adjustable to 500 ml/h; UV monitor (280 nm filter); fraction collector for 100 tubes and a recording device; a pH meter and a conductivity meter are also required
- *Buffers and chemicals*: 50 mM potassium phosphate, pH 7.0; sodium chloride for the gradients
- *Enzyme test*:

  2850 μl   100 mM potassium phosphate buffer, pH 7.5
   100 μl   oxalacetate (2 mg/ml)
    50 μl   NADH (10 mg/ml)

  3000 μl
  bring up to 30 °C and start the test with 20–50 μl enzyme preparation. Measurements are made over a period of several minutes in a 1 cm cuvette; wavelength 340 nm; coefficient of extinction = 6.3 $cm^2/\mu mol$.

Procedure

A column (1.6 cm diam. and 30 cm long) is packed with Blue Sepharose CL-6B so that the height of the column is 12 cm. After ensuring that the system is free of air, the adapter is placed on top of the column and the column is equilibrated with starting buffer (50 mM potassium phosphate buffer, pH 7.0) at a rate of 60 ml/h and a conductivity of 8 mS/cm. 20 ml of yeast suspension is made up to 40 ml with starting buffer and the pH adjusted to 7.0 and the conductivity to 8 mS/cm. The enzyme suspension is applied to the column at a flow rate of 60 ml/h and the column is then washed until OD_{280} reaches zero. The application volume and the wash volume are collected and tested for MDH activity. MDH is eluted with an increasing linear salt gradient. The salt gradient is prepared using three column volumes each of starting buffer and 'limit' buffer (starting buffer + 500 mM NaCl). The two buffer containers are linked by a bridge and the container of starting buffer is stirred with a magnetic bar. Starting with the buffer of the lower molarity the gradient is pumped onto the column at a flow rate of 60 ml/h and the fraction collector is switched on to collect fractions of about 5 ml. The elution of protein is followed on the recording device. Every fifth fraction is tested for MDH activity; in the region where MDH elutes, every fraction is measured. At the end of the run the fractions containing MDH activity are pooled and tested for MDH activity and the protein content determined. The column is regenerated by washing with one column volume of

50 mM potassium phosphate buffer, pH 7.0, containing 1 M NaCl and then equilibrating the column with starting buffer until the pH has reached 7.0 and the conductivity measures 8 mS/cm. When the results have been evaluated the experiment can be repeated using a step gradient for the elution.

Interpretation and Conclusions

A graph should be plotted of each column run; the number of fractions is plotted on the X-axis and on the Y-axis volume activity (U/ml), OD_{280}, pH and conductivity are plotted. The data before and after the affinity chromatography step should be compared in a table. The table should contain data for yield, purification factor, protein content, volume activity, total activity and specific activity. The outcome of the purification should be discussed and the two elution procedures compared.

Further reading

Affinity chromatography – principles and methods (1988) (Pharmacia).
CARLSSON, J., JANSON, J.-C. and SPARRMAN, M. (1989) Affinity chromatography. In: J.-C. Janson and L. Ryden (eds) *Protein purification – principles, high resolution methods and applications*, pp. 275–329 (Weinheim: VCH).
CLONIS, Y.D. (1990) Process affinity chromatography. In: J.A. Asenjo (ed.) *Separation processes in biotechnology*, pp. 401–46 (New York: Marcel Dekker).

4.5.4 *Purification and Characterisation of Chloramphenicol Acetyl Transferase from* Escherichia coli

M. Schweizer

Principle and Aim of the Experiment

Chloramphenicol acetyl transferase (EC 2.3.2.28; CAT) allows cells to become resistant to the normally toxic chloramphenicol because it inactivates the antibiotic by acetylating it at positions 1 and 3 in the side chain. There are several species of the enzyme found in *E. coli*, depending on the plasmid encoding the chloramphenicol resistance. The primary structure of CAT was determined by sequencing the gene and the protein encoded by it. The CAT subunit encoded by the plasmid has a molecular mass of 25 668 (Alton and Vapnek, 1979; Shaw *et al.*, 1979). The enzyme consists of four identical sub-units and catalyses the O-acetylation of chloramphenicol with acetyl-SCoA as the acyl donor according to the following equation:

$$\text{chloramphenicol} + \text{acetyl-SCoA} \rightarrow \text{chloramphenicol-3-acetate} + \text{HSCoA} \qquad (4.4)$$

A diacetyl product is also formed by CAT, but at a rate two orders of magnitude slower than the production of the monoacetyl derivative. Although the mechanism has not been explained in every detail there is some indication that the reaction is as follows:

$$\text{chloramphenicol-3-acetate} \rightleftharpoons \text{chloramphenicol-1-acetate} \qquad (4.5)$$

$$\text{chloramphenicol-1-acetate} + \text{acetyl-SCoA} \rightleftharpoons$$
$$\text{chloramphenicol-1,3-diacetate} + \text{HSCoA} \qquad (4.6)$$

The reaction shown in Eqn (4.5) is probably the result of a pH-dependent, non-enzymatic shift of the acyl residue between the positions 1 and 3 on the chloramphenicol side chain. Although the reactions given by Eqns (4.5) and (4.6) are important for the kinetic and mechanistic investigations of CAT, they in no way interfere with the enzyme test for chloramphenicol acetyl transferase which is based on the stoichiometry of the reaction given by Eqn (4.4).

Description of the Experiment and Methods

In a treatise published in 1990 Arthur Kornberg deals with the topic 'why purify enzymes?'. He gives a number of reasons for doing so and then goes on to say that no enzyme can be purified to absolute homogeneity. Specifically he writes:

> Even when other proteins constitute less than 1% of the purified protein and escape detection by our best methods, there are likely to be many millions of foreign molecules in a reaction mixture. Generally, such contaminants do not matter unless they are preponderantly of one kind and are highly active on one of the components being studied.

The purification of a protein is a prerequisite for studying its structure and function. The special properties of a protein may be exploited to purify it from the plethora of other proteins in the cell. Proteins can be separated from each other, e.g. according to their molecular weights, their solubilities or their chromatographic properties. It is absolutely essential to have a simple, user-friendly enzyme test.

Chloramphenicol acetyl transferase will be isolated from a strain of *E. coli* harbouring the plasmid pBR328 which carries the gene-encoding chloramphenicol acetyl transferase. The sequence of plasmid pBR328, which also confers resistance to ampicillin and tetracycline, is known (Soberon *et al.*, 1980).

Measurement of Chloramphenicol Acetyl Transferase in a Crude Extract of E. coli

SOLUTIONS AND EQUIPMENT

E. coli strain containing pBR328 (e.g. RR1 transformed with pBR328) (see p. 48–51); LB medium + 0.2% glucose + 50 μg/ml ampicillin; TM buffer (50 mM Tris-HCl, 0.1 mM 2-mercaptoethanol, pH 7.8); TCM buffer (TM buffer + 0.2 mM chloramphenicol); glass beads, diam. 0.45 mm (prepared by stirring for 1 h in conc. HCl and then washing with dH$_2$O to remove the acid before equilibrating with TM buffer); 5 mM acetyl-CoA (lithium salt dissolved at 20 mg/5 ml in dH$_2$O and dispensed in 0.2 ml aliquots which are stored at $-20\,°$C); DNTB 10 μM (40 mg 5,5'-dithiobis-2-nitro-benzoate in 10 ml 1 M Tris-HCl, pH 7.8); chloramphenicol 5 mM; Bradford's reagent (Bio-Rad); bovine serum albumin for preparation of the standard curve for determining the protein concentration; dialysis tubing (tubing should be boiled in 10 mM NaHCO$_3$, 1 mM EDTA and then rinsed extensively in dH$_2$O before storing at 4 °C in 1 mM EDTA); microdismembrator (Braun Melsungen) or sonifier (Branson); Amicon (for concentration of protein solutions); centrifuge tubes for JA-14 and JA-20 rotors, metal tubes for JA-20; 2 l Erlenmeyer flask, 500 ml side arm flasks (sterile); Tris-HCl 1 M, pH 7.8; ammonium sulphate, streptomycin sulphate; DE-52(Pharmacia); Sephadex G100 (Pharmacia); conductivity meter; peristaltic pump and equipment for column chromatography.

METHOD

1   5 ml LB broth (0.2% glucose + 50 μg/ml ampicillin) is inoculated with a single colony of the plasmid-containing *E. coli* strain and incubated overnight at 37 °C with shaking.

2   3 ml LB broth (as above) is inoculated with 50 μl of the overnight culture and incubated, with shaking, at 37 °C until OD_{450} = 1.2 has been reached.

3   The culture is centrifuged at 5000 rpm and 4 °C for 5 min, and the centrifugation is repeated with the supernatant. Both pellets are resuspended in 1 ml TM buffer and centrifuged again.

The following steps are performed on ice.

4   The cells are resuspended in 0.3 ml TM buffer and an equal volume of acid-washed glass beads is added. Vortex the suspension twice for 1 min (with cooling on ice in between!). The cells can also be disrupted by sonification (5 mm tip, 50 W, 2 × 20 s, with a 20 s cooling period in between) or in a dismembrator.

5   The disrupted cells are centrifuged (Eppendorf centrifuge, 12 000 rpm, 5 min, 4 °C).

6   0.2 ml supernatant is pipetted into a fresh reaction tube and placed on ice. If the enzyme test is not to be performed on the same day, the samples can be stored at −20 °C. (Under these conditions the samples can be stored for up to one month without any loss of CAT activity.)

Optical Test for the Determination of CAT Activity

The measurement of CAT activity depends on an increase in free thiol groups. The change in the concentration of the free thiol groups is determined spectrophotometrically by means of Ellman's reagent (5,5'-dithio-bis-2-nitrobenzoate = DNTB). The calculation of the results is based on a molar coefficient of extinction of 1360 (1 mol^{-1} mm^{-1}) for 5-thio-nitrobenzoic acid. The free thiol groups released upon reduction of HSCoA react with 5,5'-dithio-bis-2-nitrobenzoate to produce a 'mixed' disulphide consisting of CoA and thionitrobenzoic acid and 5-thio-2-nitrobenzoate (yellow). The conditions for measurement are:

*   wavelength 412 nm
*   light path (d = 10 mm)
*   temperature 37 °C.

The cuvettes can be either glass or plastic.

METHOD

1   A reaction mix sufficient for eight enzyme tests is prepared and placed on ice:
    | 8.8 ml | ddH$_2$O |
    | 1.0 ml | 10 μM DNTB |
    | 0.2 ml | 5 mM acetyl CoA |
    | 10.0 ml | |

2   0.588 ml of the reaction mix is pipetted into each of two cuvettes and they are allowed to equilibrate to 37 °C for 2 min.

3   20 µl of the crude extract is added to one of the cuvettes and mixed carefully. The whole is allowed to equilibrate to 37 °C for 3 min before measuring the background activity.

4   12 µl 5 mM chloramphenicol is added to start the reaction and mixed carefully; the reaction is measured for 4 min. The increase in OD_{450} with respect to time must be linear for at least 4 or 5 time points. If this is not the case the test should be repeated with a different amount of enzyme.

5   The protein concentration is determined by the method of Bradford (1976).

Calculation of CAT Activity

1   The slope is measured before and after the addition of chloramphenicol.

2   The difference between $\Delta E_{412/\text{time}}$ and $\Delta E_{412/\text{zero}}$ is determined.

3   ε_{412} for DNTB is 13.6×10^2 (l mol^{-1} mm^{-1}).

4   $d = 10$ mm; $\varepsilon d = 13.6 \times 10^3$ (l mol^{-1}) $= 13.6$ (l mmol^{-1}) $= 13.6$ (ml µmol^{-1}).

5   $V = 0.620$ ml; $v = 0.02$ ml.

A characteristic property of each enzyme is its specific activity. Enzyme activity is measured in enzyme units. The definition of an international unit is:

One unit (U) is the amount of enzyme which catalyzes the conversion of 1 µmol of substrate per minute under optimal conditions (pH, temperature, concentration of substrate, buffer).

- *Volume activity*:

$$\frac{(\Delta E/\text{min})VF}{\varepsilon dv} = \mu\text{mol}/(\text{min ml}) = \text{U/ml}$$

where V is total volume, v is sample volume, F is dilution factor.

- *Specific activity*:

$$\frac{\text{U/ml}}{\text{mg protein/ml}}$$

Purification of CAT from E. coli

The purification of CAT is performed using the method described by Shaw (1975). The steps in the purification are monitored quantitatively. Enzyme activity and the amount of protein (supernatant and pellet) are determined for each step, and the purity of the fractions is checked using SDS gel electrophoresis (see below). Detailed descriptions for ion exchange chromatography, gel filtration, setting up an elution gradient etc. are not given here, because they are to be found elsewhere in this volume. Relevant literature sources are given at the end of the subsection. Useful hints on column chromatography are to be found in the brochures 'Ion exchange chromatography: principles and methods' and 'Gel filtration: principles and methods', both available from Pharmacia.

METHOD

1   *Growth of cultures* in four 2 l Erlenmeyer flasks. Each of the four flasks containing 1 l LB broth + ampicillin is inoculated with 10 ml of a mid-log phase preculture of *E. coli* carrying the plasmid pBR328. The flasks are incubated at 37 °C with vigorous shaking.

2   *Centrifugation.* When the cells have reached stationary phase they are harvested by centrifugation (JA-14 rotor, open metal centrifuge tubes, 9000 rpm, 15 min, 4 °C) and the pellet is resuspended in 60 ml TM buffer.

3   *Cell disruption.* The cells are disrupted either in a microdismembrator or by sonication. 5 ml of the cell suspension is transferred to the sonication vessel (cooled to 4 °C) and the sonifier tip is centred correctly. Sonication is carried out for 4×1 min at level 5 (Branson sonifier) with cooling periods of 1 min between each burst of sonication. When all the suspension has been treated the disrupted cells are centrifuged (30 000 *g*, 20 min, 4 °C).

4   *Precipitation with streptomycin sulphate.* All further steps should be carried out at 4 °C. The volume of the supernatant from step 3 is determined. Streptomycin sulphate is added gradually to a final concentration of 1%. The precipitate which forms after about 30 min contains nucleic acids, and the suspension is centrifuged.

5   *Ammonium sulphate fractionation.*

   (i)   The concentration of Tris-HCl in the supernatant obtained from the streptomycin sulphate precipitation is adjusted to 100 mM with 500 mM Tris-HCl, pH 7.8.

   (ii) Finely ground ammonium sulphate is added over a period of 30 min and under constant stirring to 50% saturation (w/v). The suspension is stirred for a further 30 min before centrifuging at 30 000 *g* and 4 °C for 20 min to remove the precipitated protein. The supernatant should be poured off carefully. It should contain less than 10% of the total activity. If this is not the case, ammonium sulphate should be added as described above to 55% saturation (w/v) and the centrifugation repeated.

6   *Heat shock.* The precipitate obtained from the previous step is dissolved in TM buffer and the suspension allowed to reach room temperature before being placed in a water bath at 60 °C where it is incubated, with gentle stirring, for 10 min. After cooling, the suspension is centrifuged.

7   *Second ammonium sulphate precipitation.* Finely ground ammonium sulphate is added to the supernatant from the previous step to 50% saturation (w/v) and the suspension is stirred for a further 30 min before centrifuging. The CAT activity is to be found in the pellet from this step.

8   *Dialysis.* The pellet is dissolved in a small volume (<5 ml) of TM buffer and dialysed against TM buffer. The volume of the buffer is 3 l and it should be changed three times to ensure complete removal of ammonium sulphate.

9   *DEAE cellulose chromatography.* DEAE cellulose anion exchange chromatography is set up as follows: pre-swollen DE52 anion exchange resin is resuspended in 500 ml dH$_2$O, and when the resin has settled down the water is siphoned off carefully. This procedure is repeated twice more before the resin is placed in 0.5 M HCl for 30 min and then filtered. The acid treatment is

followed by a 30 min treatment with 0.5 M NaOH and filtration. The resin is washed with dH$_2$0 until the pH of the effluent is neutral; at this point the resin is de-gassed on a vacuum pump for 30 min. The final step is equilibration of the resin with TCM buffer.

(i) A column is packed with DE-52 to a bed volume of 2.5 × 40 cm. The column is washed with TCM buffer until the pH and the conductivity of the eluate are identical to those of TCM buffer.

(ii) The dialysed enzyme suspension is loaded onto the column and TM buffer allowed to run through the column until protein is no longer detectable at 280 nm. The protein is eluted with an increasing linear salt gradient of 0–0.4 M NaCl in TCM buffer (the volume of the salt gradient should be 1 l). CAT should be eluted between 0.15 and 0.2 M NaCl in buffer.

(iii) During the chromatography, OD$_{280}$ as measured in the flow-through cuvette is registered on the recording device. The CAT activity and the protein content is determined for each fraction. The fractions with the highest specific activity are pooled and concentrated by ultrafiltration.

10 *Gel filtration.* The enzyme solution in a volume of 5 ml is placed in the bottom of a column (2.5 × 80 cm). The column is filled with Sephadex G100 which has been equilibrated with TCM buffer and 0.2 M NaCl. Using the same buffer, the CAT preparation is eluted in an 'upward fashion'. The fractions containing CAT activity are pooled and stored at −20 °C.

Table 4.4 shows the values obtained when CAT is purified as described. The figures in the table are taken from Shaw (1975). The purified enzyme preparation can be characterised by means of the following experiments: determination of optimal substrate concentration, enzyme concentration, temperature and the dependency of the enzyme activity on pH.

Table 4.4 Purification of chloramphenicol acetyl transferase from *E. coli*.

Enzyme fraction	Total protein (mg)	Total activity (U)	Specific activity (U/mg)	Purification factor	Yield (%)
Crude extract	5400	4900	0.9	(1)	(100)
Streptomycin sulphate precipitation	4150	4650	1.1	1.2	95
1st (NH$_4$)$_2$SO$_4$ precipitation	2800	4400	1.6	1.8	90
Heat shock	900	4150	4.6	5.1	85
2nd (NH$_4$)$_2$SO$_4$ precipitation	640	4050	6.3	7.0	83
DEAE cellulose	51	2150	42	47	44
Sephadex G100	14	1750	125	139	36

CAT can also be purified by affinity chromatography as described by Zaidenzaig and Shaw (1976). This method, which relies upon the affinity of CAT for *p*-aminochloramphenicol bound to the free carboxyl group of substituted Sepharose (Sepharose NH–$(CH_2)_n$–COOH) via an acid amide formed as a result of carbodiimide activation, gives a better yield.

Further reading

ALTON, N.K. and VAPNEK, D. (1979) Nucleotide sequence of the chloramphenicol resistance transposon Tn9. *Nature* **282**, 864–9.

BERGMEYER, H.U. (1986) *Methods of enzymatic analysis*, Vol. 1: *Fundamentals*, 3rd edn (Weinheim: VCH).

BERTRAM, S. and GASSEN, H.G. (1991) *Gentechnische Methoden* (Stuttgart: Gustav Fischer Verlag).

BRADFORD, M.M. (1976) A rapid and sensitive method for the quantitation of microgram quantities of protein utilizing the principle of protein dye binding. *Anal. Biochem.* **72**, 248–54.

COOPER, T.C. (1981) *Biochemische Arbeitsmethoden* (New York: Walter de Gruyter Verlag).

EISENTHAL, R. and DANSON, M.J. (1962) *Enzyme assays* (Oxford: IRL Press).

HARRIS, E.L.V. and ANGAL, S. (1990) *Protein purification applications: a practical approach* (Oxford: IRL Press).

KORNBERG, A. (1990) In: M.P. Deutscher (ed.) Guide to protein purification. *Methods Enzymol.* **182**, 1–5.

PACKHAM, L.C. and SHAW, W.V. (1981) The use of naturally occurring hybrid variants of chloramphenicol acetyltransferase to investigate subunit contacts. *Biochem. J.* **193**, 541–52.

RODRIQUEZ, R.L. and TAIT, R. (1983) In: *Recombinant DNA techniques: an introduction*, pp. 187–92 (London: Addison-Wesley).

SCOPES, R. (1992) *Protein purification – principles and practice* (Berlin: Springer Verlag).

SHAW, W.V. (1975) Chloramphenicol acetyltransferase from chloramphenicol-resistant bacteria. *Methods Enzymol.* **43**, 737–55.

SHAW, W.V., PACKHAM, L.C., BURLEIGH, B.D., DELL, A., MORRIS, R. and HARTLEY, B.S. (1979) Primary structure of a chloramphenicol acetyltransferase specified by R plasmids. *Nature* **282**, 870–2.

SOBERON, X., COVARRUBIAS, L. and BOLIVAR, F. (1980) Construction and characterization of new cloning vehicles, IV: Deletion derivatives of pBR322 and pBR325. *Gene* **9**, 287–305.

ZAIDENZAIG, Y. and SHAW, W.V. (1976) Affinity and hydrophobic chromatography of three variants of chloramphenicol acetyltransferase specified by R factors in *Escherichia coli*. *FEBS Lett.* **62**, 266–71.

4.5.5 *SDS-polyacrylamide-gel Electrophoresis of Proteins*

M. Schweizer

Principle and Aim of the Experiment

Concentration of a protein mixture is achieved in discontinuous SDS-gel electrophoresis because the gel system used consists of a separating gel and a stacking gel, ensuring a discontinuity in the pore size of the gel matrix and the pH value. Sodium dodecylsulphate (SDS) is an aliphatic detergent consisting of twelve C atoms and a hydrophilic sulphate group. By including SDS in the gel system the proteins to be separated are immersed in SDS and acquire an evenly distributed partial negative charge because of the sulphate group. The outcome is that in a discontinuous gel the proteins are separated according to their charge and molecular weight. The relative distance travelled by the protein to be characterised can be used to determine its molecular weight because, within a particular range, the $\log_{10}$ of the molecular weights of standard proteins are proportional to the distances travelled by the proteins in the gel. For this reason the determination of the molecular weight of a protein by SDS-gel electrophoresis requires that the protein be subjected to electrophoresis in various concentrations of poly-acrylamide.

Description of the Experimental Method

The acrylamide concentration chosen depends on the molecular weights of the proteins to be separated. In general it can be said that: *the higher the molecular weight of the protein the lower the concentration of the acrylamide.* Protein standards of known size are available lyophilised from various companies (e.g. Sigma). The volume of the gel required depends obviously on the apparatus used. The values given in Tables 4.5 and 4.6 are those required for the Bio-Rad Minigel (6×8 cm $\times$ gel thickness). It is recommended that a larger volume is made up.

Table 4.5 Volumes of separating and stacking gels required for Bio-Rad Minigel.

Thickness of gel (mm)	Separating gel (ml)	Stacking gel (ml)
0.5	5.6	1.4
0.75	8.4	2.1
1.0	11.2	2.8
1.5	16.8	4.2

Table 4.6 Separation ranges of polyacrylamide gels.

Acrylamide concentration (%)	Molecular weight (kDa)
20	20–25
15	15–45
12.5	15–60
10	18–75
7.5	30–120
5	60–220

Casting the Gel and Running the Electrophoresis

MATERIAL

30% acrylamide stock solution (30% (w/v) acrylamide, 0.8% (w/v) N,N'-methylene-bis-acrylamide); the solution is filtered and then stored at 4 °C in a dark bottle. If the acrylamide used to prepare the solution is not of pa quality the acrylamide stock solution must be deionised by stirring with Dowex MR-3 resin (Sigma) at a ratio of 1 : 5 (w/v) for 30 min and then filtered through Whatman No. 1 paper. In the non-polymerised state acrylamide is a neurotoxin and can be absorbed through the skin. *GLOVES AND SAFETY MASK SHOULD BE WORN WHEN HANDLING ACRYLAMIDE!* 20% (w/v) SDS (sodium dodecyl sulphate or sodium lauryl sulphate); 10% (w/v) ammonium persulphate solution in H_2O (must be prepared fresh every time!); TEMED (N,N,N',N'-tetramethylene-ethylenediamine) (Serva); 4 × running buffer (100 mM Tris-Base, 790 mM glycine, 0.4% SDS); 1 × running buffer (250 ml 4 × running buffer + 750 ml dH_2O); 4 × buffer for the separating gel, pH 8.8 (1.5 M Tris-HCl, pH 8.8, 0.4% SDS); 4 × buffer for the stacking gel, pH 6.8 (0.5 M Tris-HCl, pH 6.8, 0.4% SDS); sample buffer (62.5 mM Tris-HCl, pH 6.8, 3% SDS (w/v), 5% β-mercaptoethanol (v/v), 10% glycerol (v/v), 0.1% bromophenol blue (w/v)) – for denaturation of proteins at least one volume must be used. (bromophenol blue solution is prepared as follows: 100 mg bromophenol blue and dH_2O are combined to a volume of 10 ml, dissolved by stirring and filtered to remove any remaining dye particles); isobutanol; 2% agarose in 1 × TBE buffer (see p. 42) to seal the glass plates before pouring the gel. Vertical electrophoresis apparatus (dimensions of the glass plates: 160 × 140 × 1.5 mm or 80 × 80 × 1 mm or 6 × 8 cm × 0.75 mm) [for Bio-Rad Minigels; one of the plates should be cut at one edge to form a shallow (ca. 2 cm deep) indentation leaving projections of about 1 cm at either side and is known as the notched plate]; spacers of appropriate width, bulldog clips; comb to form the wells (10 per gel); 8 M urea; gel dryer; Whatman 3 MM paper; cling film; 7.5% (v/v) acetic acid; Coomassie blue staining solution (0.05% Coomassie blue G-250 or R-250, 45% methanol, 9% glacial acetic acid); destaining solution (7.5% glacial acetic acid, 5% methanol); silver staining kit (Sigma); 10% glycerol.

The electrophoresis of the protein mixture is performed in vertical polyacrylamide gels. The glass plates are washed with washing-up liquid and dried with EtOH. Two grease-free glass plates separated by three spacers and held together by bulldog clips are sealed with molten agarose. Once the agarose has set, the acrylamide gel can be cast. The final gel consists of a lower separating gel and an upper stacking gel. To obtain a better separation of proteins running close to each other 8 M urea can be added to the gel. The separating gel is poured so that it reaches about 4 cm below the lower edge of the notched plate and is overlaid with isobutanol. The gel is allowed to polymerise for 1–2 h and the isobutanol is poured off before the stacking gel is poured on top of the separating gel. The protein samples can be prepared during the second period of polymerisation.

EXAMPLE OF THE METHOD

The protein samples obtained at the various steps of the purification of chloramphenicol acetyltransferase (CAT) and the purified CAT enzyme will be run in a 15%

SDS-polyacrylamide gel. Protein molecular weight standards and commercially available CAT from *E. coli* will also be subjected to the same electrophoresis. At the end of the gel run the protein bands will be visualised with either Coomassie blue or silver staining. The limit of detection with silver staining is approximately 2 ng/protein band, whereas that for Coomassie blue staining lies between 0.1 and 2 μg/protein band. Gels which have already been stained with Coomassie blue can be re-checked by silver staining.

The molecular weight of plasmid-encoded CAT can be determined by comparing the distance migrated by CAT with the distances migrated by the protein standards. The distance migrated by each of the standards is plotted against $\log_{10}$ of its molecular weight, and the molecular weight of CAT can be read from the standard curve.

METHOD

1   Dimensions of the glass plates:

    (i)   'large gel': 160 × 140 × 1.5 mm

    (ii)   'mini gel': 80 × 80 × 1 mm.

The composition and preparation of the separation gel are given in Table 4.7. The components are mixed in a vacuum flask and de-gassed with the aid of a vacuum pump.

2   0.6 ml of 2% ammonium persulphate and 10 μl TEMED are added and the gel is poured immediately. (*Caution*: the addition of these two substances initiates polymerisation of the gel matrix; therefore it is essential to work quickly.) The casting apparatus is placed in a slanting position and the separating gel is poured, avoiding bubbles, and leaving sufficient space for the stacking gel. The casting apparatus is placed in the upright position and the gel is overlaid with water-saturated butanol, added gently in a dropwise fashion, thus allowing the separating gel to polymerise without the formation of air bubbles which deform the separating gel / stacking gel interface. When the gel has polymerised (1–2 h

Table 4.7 Composition and preparation of SDS separation gels of varying acrylamide concentration.

Reagents	Gel concentration (%)			
	5 ('large')		15 ('large')	5 ('mini')
	with urea	without urea		
30% acrylamide	6.7 ml	6.7 ml	20 ml	0.8 ml
4 × separation buffer	10 ml	10 ml	10 ml	1.2 ml
Urea	19.2 g	–	–	–
H$_2$O	8 ml	22.7 ml	19.4 ml	2.85 ml
Mix and de-gas 2% NH$_4$S$_2$O$_8$	0.6 ml	0.6 ml	0.6 ml	0.122 ml
TEMED	10 μl	10 μl	10 μl	3.3 μl

for large gels, approx. 20 min for mini gels) the butanol is poured off and the interface rinsed gently with stacking gel buffer.

3   The stacking gel is now prepared according to the data given in Table 4.8.

   (i)   Stacking gel buffer (4×) = 0.5 M Tris-HCl, pH 6.8, 0.4% SDS.

   (ii)   Mix and de-gas as described for the separating gel.

   (iii) Add ammonium persulphate and TEMED and pour the separating gel to the top of the casting apparatus.

   (iv)   Insert the comb about 1 cm into the gel, shaking it gently to remove any air bubbles which may form. Allow the gel to stand for about 1 h until polymerisation is complete.

4   When the gel has polymerised, the lower, horizontal 'spacer' is removed and the gel is mounted into the running chamber. Running buffer is poured into the upper and lower chambers ensuring that in the case of the upper chamber the wells are immersed in buffer. Bubbles at the lower edge of the gel should be removed with the aid of a syringe with a bent needle.

5   The samples (1–10 μg/lane) are added to at least one volume of sample buffer, mixed and denatured by heating at 95 °C for 10 min. The samples are centrifuged briefly.

6   Prior to removing the comb, the positions of the wells are marked on the glass plate. Using both hands the comb is removed gently from the stacking gel. The wells are rinsed carefully with running buffer – this is particularly important for gels containing urea.

7   The samples are loaded into the wells, the level of buffer in the upper and lower chambers is checked and the terminals connected (top: cathode; bottom: anode). For 'large' gels, electrophoresis is carried out for 15–20 h at a constant voltage of 50–80 V and for mini gels at a constant current of 15 mA for 1.5 h. For both sizes of gel the electrophoresis is considered to be finished when the dye front has reached the bottom of the gel.

8   Switch off the power supply and disconnect the terminals before removing the gel from the running chamber. Remove the spacers from the sides of the gel and carefully pry the two glass plates apart and lift off the notched plate. If desired, the stacking gel may be trimmed off.

Table 4.8 Composition and preparation of stacking gels.

	Gel concentration (%)	
Reagents	3 ('large')	3 ('mini')
30% acrylamide	1 ml	0.25 ml
4 × stacking gel buffer	2.5 ml	0.625 ml
H_2O	6.34 ml	1.59 ml
Mix and de-gas 2% $NH_4S_2O_8$	0.25 ml	0.062 ml
TEMED	10 μl	3 μl

Fixing and Staining SDS Protein Gels

METHOD A: STAINING WITH COOMASSIE BLUE

1   The gel is stained by soaking in a solution containing 45% methanol, 9% glacial acetic acid, 0.05% Coomassie Brilliant-Blue G250 or R250 for 1 h.

2   The gel is then transferred to a destaining solution (7.5% glacial acetic acid, 5% methanol) in which, under gentle shaking at a temperature of 37 °C, the gel is destained for several hours. The destaining solution should be changed from time to time.

METHOD B: SILVER STAINING

1   Detailed instructions for silver staining are provided with the kit.

2   Gels which have already been stained with Coomassie-Blue can be re-stained with silver. It is important that the gel is thoroughly rinsed with dH_2O to remove all traces of acetate which can interfere with the silver staining.

Drying the Gels

After destaining, gels can either be stored in 7.5% acetic acid or dried.

METHOD A

1   To reduce the risk of the gel tearing during the drying process it should be placed in 10% glycerol for at least 1 h.

2   The gel is placed between Whatman 3 MM paper and cling film and laid on the gel dryer at 60 °C for 1–2 h. The length of time required for the gel to dry depends on its thickness and the efficiency of the vacuum pump connected to the dryer.

METHOD B: GEL DRYING KIT FROM PROMEGA®

1   The gel is laid in 10% glycerol for 30 min to remove any remaining traces of acetic acid.

2   One frame of the apparatus is placed in a smooth, clean tray containing dH_2O. A sheet of cellulose which has been moistened with water is placed on the frame, avoiding air bubbles. The gel is placed on the cellulose sheet, taking care to avoid air bubbles.

3   A second sheet of moistened cellulose is placed on top of the gel – no air bubbles!

4   The second frame is now laid on top and the the two cellulose sheets held together with the clips on the frames.

5   The apparatus is left in a horizontal position at room temperature for 2–3 days. If desired, the dried gel can be photographed.

Discontinuous Gel Electrophoresis under Non-denaturing Conditions

Native polyacrylamide gels will be used for the analysis of purified CAT. The composition of the gels is the same as for the Laemmli system described above, except that the running and sample buffers contain neither SDS nor β-mercaptoethanol and the acrylamide concentration used is 5%.

Example. The native conformation of CAT may be determined. Do not forget the protein standards!

Bibliography

BOLLAG, D.M. and EDELSTEIN, S.J. (1991) *Protein methods* (New York: Wiley).

HAMES, B.D. and RICKWOOD, D. (1981) *Gel electrophoresis of proteins: a practical approach* (Oxford: IRL Press).

HILLES, D.M. and MORITZ, C. (1990) *Molecular systematics* (Sunderland, MA: Sinauer Associate Publishers).

LAEMMLI, U.K. (1970) Cleavage of structural proteins during the assembly of the head of bacteriophage T4. *Nature* **227**, 680–6.

WEBER, K., PRINGLE, J.R. and OSBORN, M. (1972) Measurement of molecular weights by electrophoresis on SDS-acrylamide gels. *Methods Enzymol.* **26**, 3–27.

4.6 ISOLATION OF SECONDARY METABOLITES

U. Gräfe

4.6.1 *Basic Principles*

Antibiotics and other, generally low-molecular-weight ($\leqslant 2$ kDa), bioactive substances are produced by the wild type of microbial production strains only in relatively small amounts, e.g. $1–20 \mu g/ml$. The selective enrichment of the active substance by an appropriate method is the first step towards the isolation of the pure natural product and is essential for explaining the structure of the novel compound and the investigation of its pharmacological and therapeutic properties. In high-efficiency fermentations, often with yields of more than 40 g/l of the secondary metabolite (e.g. penicillin production by *Penicillium chrysogenum*) the aim is to harvest the active substance quantitatively from the culture media or the biomass in as few steps, ideally one, as possible, thus ensuring economical use of the biotechnological process as a whole.

In practice three methods are used: extraction of the culture media, in some cases the whole culture, with organic solvents; the selective removal of the product from the media by means of appropriate ion exchangers; or adsorbing the product on to a resin under specific, selective conditions.

General Procedure for Obtaining Secondary Metabolites from Microbial Cultures

1   Identification of the required metabolite in the biomass or in the spent medium; determination of thermal stability and the effect of hydrolysis. If it is suspected that the substance has a relatively high molecular weight, its behaviour in molecular sieves (in the form of membranes or in column chromatography) will be simulated with known substances of similar molecular weight.

2   Isolation from biological material (biomass, culture media):

  (i)   extraction of cells with methanol or the culture media with ethyl acetate or other appropriate solvents

  (ii) determination of the optimal conditions for obtaining the substance from the culture media (altering the pH and solvent; testing adsorbing and ion exchange resins and testing conditions for eluting the substance from the chosen resin)

  (iii) checking if the substance binds to acidic or basic ion exchangers

  (iv) investigating alternative methods for concentrating the substance from the fermentation medium, e.g. lyophilisation, adsorption on activated charcoal.

3   Concentration of extracts and eluates from ion exhangers: removal of the organic solvents as gently as possible ($< 60\,°C$; rotary vacuum evaporation); lyophilisation (freeze-drying) of aqueous solutions.

Extraction

Lipophilic, slightly polar and some water-insoluble metabolites are removed from spent medium by repeated extraction with organic solvents (e.g. *n*-butanol, butyl- or ethyl acetate, halogenated hydrocarbons). The solvent can be recovered by vacuum distillation after the extract has been dried. Substances still in the biomass can often be extracted with polar organic solvents, e.g. methanol or ethyl acetate. After evaporation of the solvent under vacuum the aqueous residue may be re-extracted with a apolar solvent or lyophilised by freeze-drying.

Adsorption on Ion Exchangers

Basic or acidic, strongly hydrophilic metabolites can be removed from spent medium in a single step by passage through an ion exchange column consisting of weak or strong acidic or basic resins as required, with or without complexed ions of the opposite charge. Elution of the bound secondary metabolite from the column with an acidic solution (for basic compounds), a basic solution (for acidic compounds) or salt solutions can bring about a tenfold concentration. Salt has to be removed from the solution by one of the following methods: neutralisation using an ion exchanger, chromatography on a gel matrix, dialysis or adsorption.

Adsorption on Adsorption Resin

Unspecific adsorption to resins such as Amberlite-XAD and derivatives thereof is an environment-friendly method for the concentration of organic compounds in aqueous solution and is preferable to solvent extraction. Filtration of spent medium (freed from the biomass and its pH suitably adjusted) through a column of adsorber resin removes, albeit unspecifically, most of its organophilic components. Elution of the column with a methanol–water mixture (the ratio of methanol to water can range from $7:3$ to $9.5:0.5$) allows the separation of the various organic fractions. The range of resins available also permits the separation of specific high- or low-molecular-weight substances, e.g. polypeptides. The increasing use of environment-friendly extraction and chromatography methods has led to the use of certain gases (CO_2; N_2O) rather than solvents.

Description of the Experimental Method

The development of an isolation strategy for active substances in crude concentrates is an important step in any biotechnological process and dictates the success of a screening programme for novel compounds or the commercial chances of a new therapeutically useful preparation. In the initial phases the chromatographic and other material properties of the bioactive substance are determined with a view to exploiting them in separating the substance of interest from, at least in the extraction from the wild-type strain, the high number of side products.

Strategy for the Isolation of an Unknown Low-molecular-weight Metabolite from a Crude Concentrate

- Determination of chromatographic properties with the help of thin-layer chromatography: R_f values for various types of carrier material (silica gel on glass plates, films, etc., HPTLC-, RP-, NH_2- and other modified silica gels, cellulose, polyamide) and solvent mixtures for elution. Determination of the correlation between R_f value and staining properties on the one hand and the bioactivity (antibiotic effect on test strains) on the other hand.

- Determination of specific spectroscopic properties, e.g. fluorescence or quenching thereof as well as the substance-specific staining by specific reagents (detection).

- Decision on which component corresponds to the biological activity and, depending on the results of the thin-layer chromatography, setting up preparative column chromatography with silica gel or cellulose (simple column chromatography, flash chromatography, medium-pressure chromatography, HPLC).

- Use of special chromatography methods, e.g. gel filtration on organophilic dextran gels (e.g. Sephadex type, Superdex) which allows molecular weight-dependent separation.

- Ion-exchange chromatography: use of anionic and cationic exchangers, possibly in combination with gradient elution of the adsorbed metabolite with increasing salt concentration, leading to a selective removal of undesired chemical byproducts.

- Checking the purity of the isolated product by chromatography (thin-layer chromatography [DC/HPTLC], HPLC), physicochemically by mass spectrometry (MS), nuclear magnetic resonance (NMR), light spectroscopy (UV/VIS) and biological methods (bioassay, minimal inhibitory concentration required to produce a specific effect, e.g. growth inhibition of microorganisms).

For novel substances it is the main aim to identify, on the basis of specific chemical (e.g. staining with indicator dyes), physicochemical (e.g. fluorescence, quenching of fluorescence, colour) and chromatographic (R_t values in various solvents) properties the metabolite responsible for the biological activity. At the same time the methods to be used for an optimal purification, e.g. chromatography on silica gel (normal and reverse phase), cellulose, and gel filtration on organophilic dextran gels can be determined empirically. To avoid the isolation of inactive metabolites it is essential to measure an increase in biological activity after each isolation step as manifested by the decrease in the minimal amount of the substance required to produce a defined biological effect, e.g. minimal inhibitory concentration of an antibiotic. Frequently, it is necessary to use different chromatographical methods sequentially to achieve the desired purification. Furthermore it should be noted that the use of solvents of inferior quality or of non-standardised chromatographic material can result in the contamination of partially purified preparations and may lead to the distortion of any physicochemical data obtained. Recent developments in the materials used for the solid phase in chromatography (adsorption and dispersion chromatography, gel matrices) and in the methods available (e.g. medium-pressure or flash chromatography, HPLC) for the purification of natural products provide a number of possibilities for their separation and isolation.

HPLC is the method of choice for both the preparative isolation of purified substances and the characterisation of a mixture of substances. A high degree of separation is achieved because the columns are packed with fine-grained, hydrophobic silica gels (reverse phase). The availability of special detectors (UV- and fluorescence detectors, photo-diode array detectors, refractometric detectors) permits an electrospectroscopic characterisation of all the components in the mixture as they are separated.

The purest possible preparation of the substance is essential not only for structure determination (for this purpose a maximum of 100 mg is sufficient), and the pharmacological tests (the clinical tests necessary for the approval of the substance as a drug can require up to several kilograms), but also to fulfil all the requirements of producing a substance to be used for therapeutic purposes. Physicochemical methods, e.g. mass spectrometry, nuclear magnetic resonance, UV- and visible spectrometry, refractometry, optical rotation dispersion, circular dichroism and X-ray crystallography are used for quality control and structure determination.

Structure Determination of Isolated Secondary Metabolites using Physicochemical Methods

1   *UV- and visible spectrometry.* By direct spectrophotometry of dissolved substances or DAD detection during HPLC it is possible to identify specific molecular structures (functional groups, e.g. quinones, polyenes).

2   *Refractometry.* The refractive index can be determined as a constant for each substance.

3   *Optical rotation dispersion (ORD) and circular dichroism (CD).* By measuring the angle of rotation as a function of the wavelength, it is possible to determine the absolute configuration of substituted groups in the vicinity of chromophore residues.

4   *IR spectra (IR, FT-IR).* These measurements allow the identification of $C = C$ double bonds, carbonyl and hydroxyl groups; 'fingerprint' spectra can be obtained for many substances in the range $1000-1400$ cm^{-1}.

5   *Mass spectrometry (electron pulse-induced fragmentation [EI], fast atom bombardment [FAB], field desorption [FD], electrospray [ESMS], chemical ionisation [CI]).* These methods permit the determination of molecular weight and the characteristic patterns of fragmentation, the total formula of molecules and their fragments by 'peak matching'.

6   *High-field nuclear resonance spectrometry (proton- and ^{13}C-NMR spectrometry, two- and three-dimensional correlation spectroscopy [COSY], nuclear-Overhauser-effect spectroscopy [NOE]).* With these methods it is possible to

determine the number of protons and carbon atoms in any molecule, the type and order of bonds and specific binding, information on relative configuration and the conformation of molecules.

7   *X-ray structure analysis.* The crystal structure can be determined from the X-ray diffraction patterns (relative and absolute configurations of chiral centres in the molecule).

8   *Specific chemical investigations.* The following can be carried out: qualitative chemical analysis (proportion of carbon, hydrogen, nitrogen, sulphur, etc. in substances), chemical degradation and determination of the degradation products

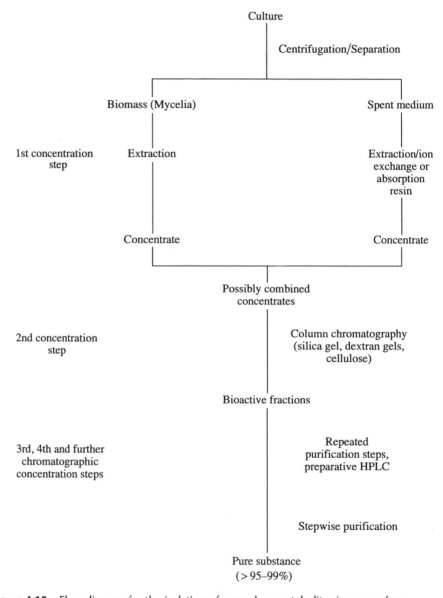

Figure 4.15   Flow diagram for the isolation of secondary metabolites in a pure form.

(e.g. by gas chromatography of volatile components), chemical synthesis of the complete and partial structures as evidence of the structure.

9   *Data bank searches.* Using the data obtained, data banks (e.g. *Chemical Abstracts*, Columbus, Ohio; Bioactive Natural Products Data Base, Szenszor, Budapest) can be searched to find any corresponding entries.

Constant quality control, as demanded by the pharmaceutical standards during the biotechnological production, excludes the possibility that any alterations in the genotype of the production strains gives rise to products with altered structures and therefore of different properties. Modern, constantly improving, analytical methods are also essential for monitoring the pharmacological properties of the novel products, for determining serum and tissue concentrations, metabolism, excretion, etc. and thereby ensuring optimal application of the drug. Several of the methods employed for the isolation of pure natural products from microbial cultures are illustrated in the following examples of solvent extraction and adsorption on ion exchangers and resins (Fig 4.15).

Comments on safety procedures. All organic solvents required for the procedures described in this section are inflammable and a health hazard. All procedures involving organic solvents must be carried out in fume cupboards. (see p. 189).

4.6.2   *Isolation of Actinomycin D as an Example for Obtaining Secondary Metabolites by Extraction*

Actinomycin D was discovered by Waksman in 1942 as a type of chromopeptide antibiotic or peptidolactone. Generally it is produced by various species of *Streptomyces* as part of a mixture of chemically similar antibiotics belonging to the actinomycin group. It contains a tricyclic hetero-aromatic phenoxazinone chromophore coupled to two pentapeptides, which contain internal cyclic esters (lactones). The phenoxazinone chromophore is responsible for the reddish-yellow coloration of the active substance. The reason for the anti-bacterial and cancerostatic effects of Actinomycin D and its homologues is the substance's ability to intercalate between the two strands of the DNA double helix. Since such substances are both carcinogenic and mutagenic it is absolutely essential to avoid skin contact and inhalation during the isolation and purification of the pure substance.

Growth of the Actinomycin D Producer *Streptomyces* spp. JA 6887

Agar slopes are inoculated with spores from the strain producing Actinomycin D, *Streptomyces* spp. JA 6887, (available from the strain collection of the Hans-Knöll-Institut für Naturstoff-Forschung e. V., Jena, Germany) and incubated for 10–14 days at 28 °C. Pieces of agar carrying surface mycelia and measuring about 3–5 cm^2 are excised and used to inoculate 500 ml straight-sided or Erlenmeyer flasks containing 50 ml preculture medium (one piece of agar/flask). The flasks are incubated at 26–28 °C on rotating tables (180 rpm) for 48 h. 3 ml of such a preculture serves as the inoculum for 50 ml culture medium contained in either 500 ml straight-sided or Erlenmeyer flasks. Incubation is carried out at 26 °C on rotating tables (180 rpm) for 96–120 h. For a further scale-up, 48-h-old samples of the main culture can be used as the inoculum for the same medium in either shake-flask cultures or in a fermenter.

- *Preculture medium*: soya extract 15 g, glucose 15 g, NaCl 5 g, $CaCO_3$ 1 g, KH_2PO_4 0.3 g; make up to 1000 ml with dH_2O and adjust the pH to 6.0 before sterilisation.
- *Culture medium*: soya extract 20 g, glucose 40 g, NaCl 5 g, $CaCO_3$ 1 g; make up to 1000 ml with dH_2O and adjust the pH to 6.5 before sterilisation.

Both media should be sterilised at 120 °C for 35 min.

Preparation of Actinomycin D by Solvent Extraction

The preparation of purified Actinomycin D from *Streptomyces* spp. JA 6887 is carried out as follows.

1 5 l of 96–120-h-old cultures (obtained by combining several cultures) is brought to pH 5.0 by the addition of acid.

2 Separation of the biomass from the medium by centrifugation or some other method.

3 The spent medium (approx. 4.5 l) is extracted twice with 1.5 l of butyl acetate or dichloromethane.

4 The biomass is extracted with 1 l of methanol and left to stand overnight. The suspension is centrifuged to remove the mycelia and the supernatant is treated in a vacuum rotary evaporator to remove the methanol. The aqueous residue is extracted twice with 500 ml butyl acetate or dichloromethane.

5 The extracts (if obtained with dichloromethane they must be dried over water-free sodium sulphate) of the spent medium and the mycelia are combined and concentrated by rotary evaporation under vacuum.

6 The addition of *n*-hexane or petroleum ether (bp 30–50 °C) leads to the precipitation of crude Actinomycin D. The precipitation is complete after being left to stand overnight in a cold room. The upper liquid phase is decanted and filtered.

7 The precipitated crude preparation of Actinomycin D is dissolved in a few ml toluene. The solution is passed through a column of silica gel 60 (0.063 to 0.1 mm) in toluene and eluted with a mixture of toluene and ethyl acetate (2:8, v/v). The fractions are tested for bioactivity and their purity is checked by thin-layer chromatography on silica gel C (Merck): liquid phase: benzene–ethylacetate–methanol (10:2.5:1 v/v; R_f ca. 0.24). An alternative method is circular filter chromatography using the system *n*-dibutyl-ether/*n*-butanol (1:5; saturated with a 2% aqueous solution of sodium naphthyl sulphonate). The liquid phase and the saturating solution are mixed in equal proportions and the upper phase is used to impregnate the chromatography paper, which has to be dried before use.

8 The fractions containing Actinomycin D are concentrated in a rotary vacuum concentrator. The red, possibly crystalline substance obtained (Actinomycin D) is dried in an exsiccator over calcium carbonate.

4.6.3 *Isolation of Nourseothricin from a Culture of* Streptomyces noursei *JA 3890b – an Example of the Use of Ion-exchange Chromatography*

Nourseothricin is a virtually equimolar mixture of Streptothricin F and D containing one and three lysyl residues, respectively, and traces of Streptothricins E and C.

Streptothricins are relatively common broad-spectrum streptomycete antibiotics which are effective against Gram-positive and Gram-negative bacteria, fungi and various viruses. However, they have little or no application in the treatment of infectious diseases because of their toxicity. Nevertheless the fact that they are poorly absorbed by the gastro-intestinal tract means that Streptothricins are used in animal husbandry for promoting growth.

These antibiotics are highly water soluble and, as such, cannot be obtained from cultures by extraction or by absorption onto resins. The use of weakly acidic cation exchangers (e.g. Wofatit CP, Amberlite resins) is therefore the method of choice for the isolation of basic Streptothricins and the related aminoglycosides (e.g. Streptomycin, Neomycin and Gentamycin) from cultures producing them. The following procedure describes the isolation of Nourseothricin and the detection of its components by bioautographic and chromatographic reagents. Substances purified by this method may contain salts, and the biological activity is always determined in relation to the pure product as the standard.

Cultivation of the Nourseothricin-producing Strain *Streptomyces noursei* JA 3890b

Streptomyces noursei JA 3890b (strain collection of the Hans-Knöll-Institut für Naturstoff-Forschung e. V., Jena) or a mutant or variant thereof is propagated as an agar slope. As described in the previous section, agar pieces $3-5$ cm^2 in size are used to inoculate 50 ml of nutrient medium in 500 ml straight-sided or Erlenmeyer flasks which are incubated at 28 °C on rotating tables (180 rpm). $3-5$ ml of a 48-h-old culture is the inoculum for 50 ml of nutrient medium or complex medium (in either straight-sided or Erlenmeyer flasks) to obtain the main culture. Incubation is carried out under the same conditions as those described for the preculture. A further scale-up is possible with a 48-h-old culture being used as a 10% inoculum for a shaken-flask or fermenter culture. At the end of the incubation period the biomass is separated from the media by centrifugation. The spent medium is then subjected to ion-exchange chromatography.

- *Preculture medium*: glucose 15 g, soya extract 15 g, NaCl 5 g, CaCO$_3$ 1 g, KH$_2$PO$_4$ 0.3 g. Make up to 1000 ml with dH$_2$O and adjust the pH to 6.5 before sterilising.
- *Culture medium*: corn starch 80 g, glucose 5 g, (NH$_4$)$_2$SO$_4$ 6.5 g, KH$_2$PO$_4$ 0.08 g, NaCl 2.5 g, KCl 2.5 g, MgSO$_4 \times$ 7 H$_2$O 0.25 g, MnSO$_4 \times$ 2 H$_2$O 0.02 g, FeCl$_3 \times$ 6 H$_2$O 0.005 g, CaCO$_3$ 1 g. Make up to 1000 ml with dH$_2$O and adjust the pH to 6.5 before sterilisation.
- *Complex medium*: potato starch 32 g, glucose 29 g, soya extract 14 g, NH$_4$NO$_3$ 7 g, MgSO$_4 \times$ 7 H$_2$O 0.25 g, NaCl 1 g, CaCO$_3$ 6 g. Make up to 1000 ml with dH$_2$O and adjust the pH to 6.0 before sterilisation.

All media are sterilised at 120 °C for 35 min.

Isolation of Nourseothricin from the Culture Medium of *S. noursei* JA 3890b

The isolation of the basic, water-soluble antibiotic is performed as follows.

1   5 l of a 96-h-old culture is centrifuged or filtered to remove the mycelia which are then discarded.

2   *Adsorption on a weakly acidic cation exchanger.* 4.5 l of the clear culture solution is loaded onto a column (80 cm × 3 cm) packed with a weakly acidic (carboxyl ions-) cation exchanger in the sodium state (e.g. Wofatit CP, Amberlite IRC-50).

3   *Elution from the ion exchange column.* The remains of the culture medium are washed out with water before eluting Nourseothricin with 0.1 N HCl and collecting fractions of 20 ml volume. The fractions are tested for anti-bacterial activity against the tester strain, *Bacillus subtilis* ATCC 6633. Those fractions containing Nourseothricin are neutralised by the addition of a weakly basic anion exchanger resin (e.g. Wofatit L-150 or Amberlite IRA-68 or similar; −OH form), combined and filtered to remove the ion exchanger before evaporating to almost dryness at 60 °C under vacuum.

4   *Gel filtration chromatography.* 10 ml methanol is added to the aqueous residue (ca.10 ml), and any salt which precipitates out is removed by filtration. The solution is loaded onto a column (60 cm × 3.5 cm) packed with Sephadex LH-20 (in methanol/H_2O; 9 : 1) and eluted with the same methanol/water mixture. Fractions of 15 ml are collected.

5   *Nourseothricin hydrosulphate.* The antibiotic activity of the fractions is tested using the tester strain ATCC 6633, and those with antibiotic activity are combined. These fractions are acidified by the addition of H_2SO_4 (pH 5.0), and the aqueous solution is evaporated to dryness under vacuum at a temperature maximum of 60 °C or lyophilised by freeze-drying.

6   The preparation is de-salted by gel filtration on dextran gels (e.g. Sephadex G-25, G-50 or G-75 and eluting with water.

7   *Chromatographic testing of isolated Nourseothricin.* 50 μl of the Nourseothricin solution is pipetted onto circles of chromatography paper and allowed to dry naturally. The chromatography is run overnight with *n*-propanol−pyridine−glacial acetic acid−water (15 : 10 : 3 : 12; w/v). The solvent is allowed to evaporate before spraying the chromatogram with 5% ninhydrine in ethanol / glacial acetic acid (8 : 2; v/v). Concentric zones corresponding to Nourseothricin components F, E, D and possibly C (in the order of magnitude of their R_f values) appear, and their anti-bacterial activity is tested by parallel bioautography. The upper phase of a mixture of chloroform/methanol / 17% aqueous ammonia acts as the running phase for thin-layer chromatography on silica gel plates (R_f Streptothricin 0.26).

4.6.4   *Isolation of Turimycin from the Culture Supernatant of* Streptomyces hygroscopicus *JA 6599 using an Adsorber Resin*

Turimycin consists of a mixture of several, chemically very similar, macrolid antibiotics of the leucomycin group. These macrolides are produced by *Streptomyces hygroscopicus* JA 6599 (strain collection of the Hans-Knöll Institut für Naturstoff-Forschung e. V., Jena) and excreted into the medium. In general, macrolides are weakly toxic, lipophilic antibiotics containing an aglycone and several sugar residues. Several members of this class of substances, e.g. Erythromycin, Olean-domycin and Tylosin, are widely used as broad-spectrum antibiotics.

At the end of the fermentation the pH of the culture supernatant is adjusted to slightly basic before Turimycin is extracted with an organic solvent, e.g. butyl acetate. An alternative method, described here, is the adsorption of the culture supernatant onto an adsorption resin.

Growth of *Streptomyces hygroscopicus* JA 6599

Starting with conserved spores of *Streptomyces hygroscopicus* JA 6599, agar slants are obtained after 10 days incubation at 25 °C. 50 ml of preculture media in 500 ml straight-sided or Erlenmeyer flasks are inoculated with $3-5$ cm^2 slabs of the agar slants and incubated on a rotating table (180 rpm) for 48 h at 25 °C. 3 ml aliquots of these cultures are used to inoculate 50 ml culture media in 500 ml straight-sided or Erlenmeyer flasks which are incubated at 25 °C and 180 rpm for 96 h. A further scale-up is possible using the same conditions as described for *Streptomyces noursei* in Subsection 4.6.3.

At the end of the fermentation the pH of the culture is brought to 5.0 by the addition of dilute sulphuric acid before separating the mycelia from the culture supernatant.

- *Preculture medium*: glucose 15 g, corn steep liquor (dry weight) 10 g, dried yeast 5 g, CaCO$_3$ 4 g. Make up to 1000 ml with dH$_2$O and adjust the pH to 6.9 before sterilisation.

- *Culture medium*: potato starch 40 g, glucose 5 g, molasses 5 g, soya flour 20 g, dried yeast 3 g, CaCO$_3$ 2 g. Make up to 1000 ml with dH$_2$O and adjust the pH to 6.8 prior to sterilisation. Medium should be sterilised at 120 °C for 35 min.

Isolation of Turimycin

1 *Adsorption on the adsorber resin*. The supernatant (ca. 5 l; 100–1000 mg Turimycin/l) of a 96 h culture is brought to pH 8.0 by the addition of sodium hydroxide and applied to a column (80 cm × 5 cm) packed with adsorber resin (Amberlite XAD 1180 or XAD 16, equilibrated with water). The rate of flow through the resin is 5 ml/min. When all the culture supernatant has been loaded, the column is washed with water.

2 *Elution of the resin*. The antibiotic is eluted from the resin with 2 l methanol/ water (8 : 2; v/v) and 20 ml fractions are collected. The antibiotic content of the fractions is determined by their anti-bacterial effect on the tester strain *Bacillus subtilis* ATCC 6633, and the turimycin-containing fractions are pooled. Methanol is removed by vacuum distillation, and the pH of the aqueous residue is brought to 8.0 by the addition of 1 M NaOH. The aqueous solution is extracted with three 150 ml aliquots of ethyl acetate.

3 *Crude Turimycin*. The extract is dried over sodium sulphate and the solvent is removed in a rotary vacuum evaporator to produce a colourless or brownish substance which is further purified by chromatography.

4 *Chromatographic purification of the turimycin complex*. A suitable system is column chromatography using silica gel 60 (0.063 to 0.1 mm) and toluene/ acetone (5 : 3; v/v) or dichloromethane/methanol (95 : 5; v/v) as the eluant, which results in a partial separation of the component substances. A further

possibility is gel filtration chromatography on organophilic dextran gels, e.g. Sephadex LH-20 with methanol as the eluant.

5   The presence of Turimycin can be demonstrated by thin-layer chromatography (silica gel plates; benzene/acetone; 5:3; v/v); R_f 0.06 to 0.18; visualisation of Turimycin is achieved by spraying with 5% vanillin and conc. H_2SO_4) or bioautographically (strips of the chromatogram are laid on a plate inoculated with *Bacillus subtilis* ATCC 6633).

Further reading

Isolation of Secondary Metabolites from Microbial Cultures

PAPE, H., REHM, H.-J. and REED, G. (eds) (1986) *Biotechnology*, Vol. 4, (Weinheim: VCH).

VANDAMME, E.J. (ed.) (1984) *Biotechnology of industrial antibiotics* (New York: Marcel Dekker).

VERALL, M.S. (ed.) (1985) *Discovery and isolation of microbial products* (Chichester: Ellis Horwood).

Chromatographic Detection of Antibiotics

JORK, H., FUNK, W., FISCHER, W. and WIMMER, H. (1989) *Dünnschicht-Chromatographie* (Weinheim: VCH).

LAATSCH, H. (1988) *Die Technik der organischen Trennanalyse* (Stuttgart: Georg Thieme Verlag).

RANDERATH, K. (1965) *Dünnschicht-Chromatographie* (Weinheim: VCH).

WAGMAN, G.H. and WEINSTEIN, G. (1984) *Chromatography of antibiotics* (Amsterdam: Elsevier).

General References on Antibiotics

BYCROFT, J. (1980) *Dictionary of antibiotics* (London: Chapman and Hall).

GRÄFE, U. (1992) *Biochemie der Antibiotika* (Heidelberg: Spektrum).

GRÖGER, D. and JOHNE, S. (1982) *Mikrobielle Gewinnung von Arzneistoffen – Pharmazeutische Mikrobiologie* (Berlin: Akademie Verlag).

LANCINI, G. and PARENTI, F. (1982) *Antibiotics: an integrated view* (Berlin: Springer Verlag).

ONKEN, D. (1983) *Antibiotika – Chemie und Anwendung* (Berlin: Akademie Verlag).

VINING, L.C. (1983) *Biochemistry and genetic regulation of commercially important antibiotics* (London: Addison-Wesley).

WAGMAN, G.H. and COOPER, R. (eds) (1989) Natural products isolation. Separation methods for antimicrobials, antivirals and enzyme inhibitors. *J. Chromatogr.* Library, Vol. 43, (Amsterdam: Elsevier).

Special References for the Preparative Methods Described

Nourseothricin

BOCKER, H. and BERGTER, F. (1986) Nourseothricin – Eigenschaften, Biosynthese, Herstellung. *Arch. Exper. Vet. Med.* **40**, 646–57.

BORDERS, D.B., SAX, K.L., LANCASTER, J.E., HAUSSMANN, W.K., MITSCHER,

L.A., WETZEL, E.R. and PATTERSON, E.L. (1970) Structures of LL-AC541 and LL-AB644, new streptothricin-type antibiotics. *Tetrahedron* **26**, 3123–33.

Turimycin

FRICKE, H., SCHADE, W. and RADICS, L. (1984) Isolation, characterization and structure elucidation of 2',9',18'-tri-O-acetyl-3,18-hemiacetyl-turimycin H3. *Pharmazie* **39**, 414–16.

GERSCH, D., BOCKER, H. and THRUM, H. (1977) Biosynthetic studies on the macrolide antibiotic turimycin using ^{14}C-labelled precursors. *J. Antibiot.* **30**, 488–93.

HAUPT, I., FRICKE, H., CERNA, J. and RYCHLIK, J. (1976) Effect of leucomycin-like macrolide antibiotic turimycin on ribosomal peptidyltransferase from *Escherichia coli*. *J. Antibiot.* **29**, 1314–19.

Actinomycin

BROCKMANN, H. and MANGOLD, J.H. (1964) Partialsynthese von Actinomycin $C_1(D)$. *Naturwissenschaften* **51**, 383–4.

Safety Notes (see p. 183)

It is an incorrect assumption that all natural products are harmless and cause no damage! For instance, Actinomycin D and other actinomycins are cytotoxic and mutagenic. On account of these properties, Actinomycin C has been used as a chemotherapeutic agent. Skin contact and inhalation of actinomycins has to be avoided at all costs. All solvent residues, filter papers, etc. which have been in contact with actinomycins must be treated as special waste and are to be disposed of as such. However, by taking the precautionary measures referred to above, the risk involved in working with actinomycins is no greater than that encountered when working in a chemistry laboratory.

5

Special Techniques

5.1 MICROBIAL SENSORS

K. Riedel

Basic Principles

Microbial sensors, a combination of immobilised microorganisms and physical transducers, bring together the specificity of biological systems and the high sensitivity of physical measurement. They transform the biochemical information of a substrate into a physically quantifiable, preferably electrical, signal which can then be amplified electronically. The transducer can be potentiometric (pH electrodes, ion selective electrodes) or amperometric (oxygen electrode, optoelectronic detectors, thermistors or field effect transistors) electrodes. The choice of transducer is determined by the metabolic process selected for the measurement. The mechanism underlying microbial sensors can be explained as follows. The substance or mixture to be determined is taken up by the microorganism, metabolised within the cell, and the metabolites are excreted. Oxygen is required under aerobic conditions. The metabolites and the oxygen concentration can be measured amperometrically (oxygen) or potentiometrically (pH, NH_4^+, CO_2, H_2S). By far the most commonly used transducer in microbial sensors is the amperometric oxygen electrode.

The application of microbial sensors is determined by their specificity:

- multireceptor activity, i.e. these sensors are characterised by their wide range of applications
- variability, or the ability to adjust to new requirements
- stability (the stability of microbiological sensors is due to their ability to regenerate themselves much better than enzyme sensors do)
- economical production costs; no complicated enzyme preparations.

These properties, in particular the multireceptor activity, make microbiological sensors ideal for determining multiple parameters, e.g. biochemical oxygen requirement, toxicity or mutagenicity. The lack of substrate specificity means that microbiological sensors are not suitable for substrate determinations. They can, however, be used successfully for determining pure substrates, if enzymes are not available or unstable, or when a cofactor must be regenerated as in the determination of aromatic substances (phenol, benzoate and their chlorinated derivatives).

The measurement principle of microbiological sensors is shown schematically in Fig. 5.1. Oxygen diffuses from the culture being measured, through the dialysis membrane. The microorganisms on the electrode use some of the oxygen; i.e respiration results in a reduction of the oxygen tension and is illustrated by the steady-state current I_E (Fig. 5.1(2)). A substance or mixture of substances, e.g a sample of effluent, added to the culture is taken up by the microorganisms and the respiration is increased. The oxygen tension falls, resulting in a reduced current. The current falls until a new steady-state I_S is reached (Fig. 5.1 (3)). The difference between the steady-state currents I_E and I_S reflects the respiration rate for the added substrate. The alteration in the respiration rate after substrate addition, termed the acceleration of respiration, is measured as the maximum alteration in current per time unit.

There are two ways of taking measurements with a microbial sensor:

1 measurement after a specific time, i.e. measuring the change in the current, or

2 determination of the rate of change in the current after addition of the sample.

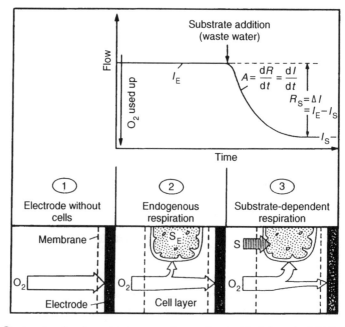

Figure 5.1 Oxygen tension in various sensors (according to Riedel *et al.*, 1990).

Construction of a Microbial Sensor

The principle of setting up a microbial sensor is shown schematically in Figs 5.2 and 5.3.

1   The microbial strain to be used as the biosensor is grown in the appropriate manner.

2   The cells are harvested by centrifugation and the cell pellet is resuspended in 0.1 M phosphate buffer, pH 6.8 at a concentration of 0.150 to 0.500 g wet weight/ml. (Wet weight is approximately four times dry weight.)

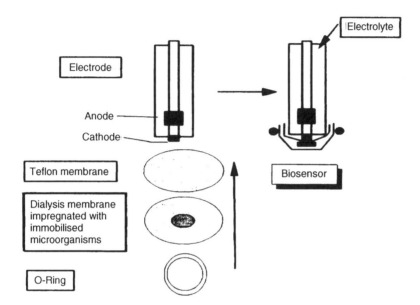

Figure 5.2   Construction of a biosensor.

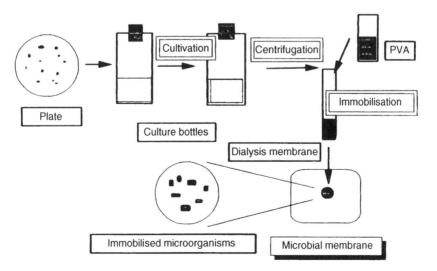

Figure 5.3   Production and immobilisation of microorganisms for biosensor construction.

3   The cell suspension is mixed with an equal volume of a 5% solution of polyvinyl alcohol (PVA). The use of a vortex mixer ensures that the cells are distributed homogeneously.

4   20 μl of the PVA cell suspension is transferred (by spotting) to a dialysis membrane with a pore diameter of 0.6 μm.

5   The membrane is stored at 4 °C for 24 h.

6   The dried, microbial-impregnated membrane is placed on the Teflon membrane of an oxygen electrode so that the microbial 'spot' is aligned above the cathode. The membrane is held in place with an O-ring. Another possibility is to lay the cell-impregnated dialysis membrane on top of the Teflon membrane and then fix the two membranes on top of the oxygen electrode with the aid of an O-ring. The electrode is filled with a saturated solution of KCl which serves as the electrolyte and then connected to the measuring and recording devices.

7   The microbial sensor is placed in a 20 ml glass beaker containing 10 ml buffer, and this constitutes the measuring cell. This is stirred constantly to maintain the oxygen supply, and the temperature is kept at 37 °C or room temperature depending on the requirements of the culture used.

8   Newly prepared microbial sensors usually require 2–12 h until they are 'activated'; the time required depends on the concentration of the cell suspension used to produce the sensor; the higher the cell concentration the longer the time required for activation. The activation of the biosensor can be checked by measuring the base current and testing the effect of the addition of the substance whose effect is to be determined.

9   Measurements are carried out after the addition of the substrate (usually in a volume between 50 and 200 μl). Addition of the substrate must be done rapidly; slow addition of the substrate may cause too low a signal. As soon as the measurement has been made, the electrode must be rinsed.

10 The effect observed can be calculated in one of the following ways:

  (i)   If the effect was measured constantly on a recording device the maximal rate of the increase of the current with respect to time can be determined graphically (nA/min).

  (ii) Alternatively the difference in current before and after the addition of the substrate, once it has reached a maximum again, can be measured.

  (iii) *Calculation of the biomass load (L) of the sensor membrane*:

$$L = \frac{WW[\text{mg/ml}]V[\text{ml}]}{4r^2[\text{cm}]} \times [\text{mg}\,DW/\text{cm}^2]$$

  where WW is wet weight of the cell suspension (before immobilisation); V is volume of the PVA suspension applied to the membrane; r is radius of the biomass 'spot'.

Examples

Glucose Determination with *Candida spp.*

1   *Cultivation. Candida* spp. are grown under aerobic conditions (shaking; 50 ml

culture volume in a 500 ml flask) for 24 h at 25 °C in 0.3% malt extract, 0.3% peptone, 0.3% yeast extract and 1% glucose.

2   *Immobilisation*. As described above.

3   *Measurement*. The effect of the glucose concentration on the glucose response is determined by testing solutions in the range 0.1–10 mmol (final concentration in the electrode).

The substrate spectrum is determined by testing various sugars, e.g. saccharose, maltose, fructose, etc.

Determination of Aromatic Compounds with *Rhodococcus spp.*

1   *Cultivation. Rhodococcus* spp. are grown under the aerobic conditions described above for *Candida* spp. for 48 h at 25 °C. The medium contains: 0.9% $Na_2HPO_4 \times 12\ H_2O$, 0.15% K_2HPO_4, 0.1% NH_4Cl, 0.02% $MgSO_4 \times 7\ H_2O$, 0.002% $CaCO_3$, 0.001% $FeSO_4 \times 7\ H_2O$, 50 $\mu g/l$ H_3BO_4, 10 $\mu g/l$ $CuSO_4 \times 5\ H_2O$, 10 $\mu g/l$ KI, 40 $\mu g/l$ $ZnSO_4 \times 7\ H_2O$, 24 $\mu g/l$ $Na_2MoO_4 \times 2\ H_2O$. After 24 h of growth, phenol is added to a final concentration of 0.5 g/l.

2   *Immobilisation*. See above.

3   *Measurement*. The dose-response effect of phenol is determined in the range 0.1–10 mmol (final concentration in the electrode).

The range of substrates is determined by testing various aromatic compounds, e.g. benzoate, halogenated phenols.

References

RIEDEL, K., LANGE, K.-P., STEIN, H.J., KÜHN, M., OTT, P. and SCHELLER, F. (1990) A microbial sensor for BOD. *Water Res.* **24**, 883–7.

RIEDEL, K., RENNEBERG, R., WOLLENBERGER, U., KAISER, G. and SCHELLER, F. (1989) Microbial sensors: Fundamentals and applications for process control. *J. Chem. Tech. Biotechnol.* **44**, 85–106.

SCHELLER, F. and SCHMID, R.D. (1992) *Biosensors: fundamentals, technologies and applications*. GBF Monographs, Vol. 17 (Weinheim: VCH).

5.2   OUCHTERLONY IMMUNO-DOUBLE DIFFUSION TEST

M. Schweizer

The immuno-double diffusion test is a simple and effective method for determining the titre of antisera prepared against purified enzymes.

Materials

Borate/saline buffer (5 vols 0.1 M Na borate, pH 8.4, 95 vols physiological saline [0.9 % NaCl]); (preparation of 0.1 M Na borate buffer: 6.2 g boric acid, 9.5 g borax and 4.4 g NaCl are dissolved in 1000 ml dH$_2$O); agarose gel (0.175 g agarose in 25 ml borate/saline buffer); glass plates (6 × 6 cm); NaN$_3$; punch for making wells in the agarose; antigen, antiserum.

Procedure

1 Agarose-coated glass plates are prepared as follows: 0.175 g agarose is dissolved by heating in 25 ml borate/saline buffer to which a pinch of NaN$_3$ has been added. 6 ml molten agarose is pipetted carefully onto each of the glass plates. The plates are left to solidify at room temperature for at least 30 min. The plates can be stored at 4 °C for several days until required, by placing them in a petri dish lined with moistened filter paper.

2 In preparation for the test, wells are punched in the agarose in the following pattern: one central well surrounded by six wells arranged hexagonally.

3 Antigen (40 μl, approx. 10–40 μg) is pipetted into the central well and serial dilutions of antiserum in physiological saline are pipetted into the peripheral wells. The plates are allowed to stand at room temperature for up to 3 days to allow the formation of the precipitation lines by antigen/antiserum interaction.

A visible immune reaction is represented by a white precipitation line between the central and the peripheral wells. The titre of the antiserum is the highest dilution of the antiserum which reacts with the antigen to give a precipitation line.

Reference

OUCHTERLONY, Ö. (1962) Diffusion-in-gel methods for immunological analysis, II. In: P. Kallos and B.H. Waksman (eds) *Progress in allergy*, Vol. VI, pp. 30–154 (Basel: Karger).

5.3 TRANSFER OF PROTEINS TO NITROCELLULOSE (WESTERN BLOTTING)

M. Schweizer

Principle of the Experiment

The technique described here is analogous to that developed by E. Southern for the transfer of DNA fragments from an agarose gel onto a nitrocellulose membrane (Southern blotting). Western blotting is the transfer of proteins which have been subjected to electrophoresis in polyacrylamide from the gel matrix onto a nitrocellulose membrane. The proteins which have been transferred can be identified either histologically or immunologically. The antigen–antibody complex can be detected directly or indirectly. Protein A is used for the *direct detection* of the antigen–antibody complex. Protein A can be isolated from the cell wall of *Staphylococcus aureus*; it has a molecular mass of 42 kDa and a high affinity for the invariable part (F$_c$) of the antibody. Protein A can be coupled to horseradish peroxidase, alkaline phosphatase, biotin or gold particles. Furthermore the antigen–antibody reaction is not affected. In the *indirect method* the bound specific antibody is allowed to react with a secondary antibody which has been labelled to allow its detection.

The immunocolour reaction with protein A / peroxidase conjugate (sensitivity: 10 pg for peroxidase or alkaline phosphatase and 1 pg for immunogold labelling) renders filter-bound antigen–antibody complexes visible. After the proteins have been transferred from the gel to the membrane, the membrane is incubated in antiserum raised against the protein being tested. The antigen–antibody complexes formed during the incubation are then visualised with an ELISA (enzyme-linked immuno-sorbent-assay) test system based on protein A conjugated to horseradish peroxidase or alkaline phosphatase. Chromophor substrates for horseradish peroxidase and H_2O_2 are either 3,3'-di-aminobenzidene or 4-chloro-1-naphthol and, for alkaline phosphatase, nitro-blue tetrazolium or 5-bromo-4-chloro-3-indolylphosphate.

A kit is available from Amersham which allows antigen–antibody complexes linked to either horseradish peroxidase or alkaline phosphatase to be detected as chemiluminescence. Peroxidase catalyses the oxidation of luminol, a cyclic diacylhydrazide, in the presence of H_2O_2 to 3-aminophthalate and light with a wavelength of 428 nm. The light emitted by this reaction can be increased with the aid of an enhancer.

The substrate for the chemiluminescent reaction with alkaline phosphatase is (3-(2'-spiroadamantan)-4-methoxy-4-(3'-phosphoryloxy)-phenyl-1,2-dioxetan), a derivative of adamantan which is dephosphorylated to an unstable intermediate and then to adamantanon and a phenolate anion with the simultaneous emission of light. Both chemiluminescent detection systems have a high sensitivity. The chemiluminescence emitted is detected on an X-ray film. Similar kits are available from other companies.

Description of the Experiment

It is possible to make up the solutions required and so avoid the expense of a kit. The reagents 3,3'-diaminobenzidine (*carcinogenic*) and protein A / horseradish peroxidase are available from Sigma. Protein A binds specifically to the constant region (F_C) of immunoglobulins, and the peroxidase transfers electrons from diaminobenzidine to hydrogen peroxide. As a result of the oxidation, diaminobenzidine is reduced to a brown insoluble substance and gives rise to the signal on the nitrocellulose membrane thus marking the position of the antigen–antibody reaction. The intensity of the signal can be increased by the addition of cobalt and nickel ions, giving a final signal which is dark blue in colour.

Material

Transfer buffer (39 mM glycine, 48 mM Tris-base, 0.0375% (w/v) SDS, 20% (v/v) methanol); 1 × PBS (10 mM Na phosphate, pH 7.2, 150 mM NaCl); 1 × TBS (10 mM Tris-HCl, pH 7.5, 150 mM NaCl) (PBS may be used instead of TBS); 1 × TBST (1 × TBS + 0.05% (v/v) Tween 20); blocking solution BLOTTO = bovine lacto transfer technique optimiser (5% (w/v) skimmed milk powder in 1 × TPBS; 0.01% Thiamersol (or a few crystals of NaN_3) or 3% (w/v) gelatine in 1 × TPBS reduces the background coloration of the filter). Specific antiserum against the protein being tested: the antiserum can be diluted from 1 : 10 to 1 : 100 000 in either 0.5% (w/v) BSA (bovine serum albumin) / 1 × TPBS, 0.01% (v/v) Thiamersol or 1% (w/v) gelatine / 1 × TPBS, 1% (v/v) Thiamersol. The titre of the antibody should be determined. The antibody solution can be re-used several times. 50 mM K phosphate buffer, pH 7.5 (50 mM K_2HPO_4, 50 mM KH_2PO_4; 50 mM KH_2PO_4 (approx. 300 ml) is adjusted to pH 7.5 (room temperature) by the addition of approx. 2 l of 50 mM K_2HPO_4; the pH must be checked with a pH meter. Protein A / horseradish peroxidase conjugate (Sigma) (0.5 mg conjugate is dissolved in 1 ml ddH_2O and dispensed in 12.5 μl aliquots to be stored at $-20\,°C$). Before use, an aliquot of the conjugate is added to 50 ml antiserum dilution (end concentration of the conjugate is 1 in 50 000); 30% H_2O_2 (v/v). Electroblotting apparatus (e.g. Pharmacia Multiphor II Novablot); power supply with constant voltage and current; plastic trays; Whatman 3MM paper; nitrocellulose 0.45 μm; cling film; etc.

ELISA Staining Reagents

FOR HORSERADISH PEROXIDASE

- Diaminobenzidine staining solution (to be made up immediately before use: 60 mg diaminobenzidine tetrahydrochloride (Sigma) is dissolved in 97 ml 50 mM K phosphate buffer, pH 7.5. (*Caution!* carcinogenic; appropriate precautions should be taken.) 1.5 ml each of 1% $CoCl_2$ and 1% $Ni(NH_4)_2(SO_4)_2$ are added under stirring.

- *Alternatively*: diaminobenzidine can be replaced by 4-chloro-1-naphthol (Sigma). 100 mg 4-chloro-1-naphthol is dissolved in 1 ml methanol. To prepare 50 ml staining solution, 250 μl 4-chloro-1-naphthol is added to 10 ml methanol and 40 ml 1 × TBS, and the reaction is started by the addition of 175 μl H_2O_2.

FOR ALKALINE PHOSPHATASE

p-nitro-blue-tetrazolium (NBT) solution (50 mg/ml NBT in 70% (v/v) dimethyl formamide); 5-bromo-4-chloro-3-indolyl-phosphate-*p*-toluidinium salt (BCIP) solution (50 mg/ml BCIP in dimethyl formamide; these solutions can be dispensed in aliquots for storage at $-20\,°C$). To prepare the alkaline phosphatase reagent, add 330 μl NBT and 170 μl BCIP to 50 ml 100 mM Tris-HCl, pH 9.5, 100 mM NaCl, 5 mM $MgCl_2$ immediately prior to use.

Electroblotting of Proteins

Proteins can be transferred from polyacrylamide gels to nitrocellulose by means of electroblotting: i.e. the transfer of the protein is accelerated by means of electric current and takes no longer than 3 hours!

Method

1   The graphite anode of the electroblotter is moistened with transfer buffer. Ten layers of Whatman 3MM paper, cut to the dimensions of the gel and presoaked in

transfer buffer, are placed on the anode. Each sheet of paper is 'rolled out' with a test tube or pipette to remove any air bubbles. A sheet of nitrocellulose with the dimensions of the gel and moistened with transfer buffer is placed on the stack of filter paper. The gel is placed on the nitrocellulose, and any air bubbles are removed before another ten sheets of moistened Whatman 3MM are laid on top.

2 The graphite cathode is placed on top of the paper/gel sandwich. A current of 0.8 mA/cm^2 is applied for 3 h. The voltage should not exceed 40 V.

Immunostaining with Protein A / Horseradish Peroxidase Conjugate

For explanatory purposes both the direct (3,3'-diaminobenzidine and protein A horseradish peroxidase conjugate) and the indirect (NBT, BCIP and anti-human-IgG / alkaline phosphatase) methods will be described.

Direct Method

1 The electroblot is dismantled and the gel stained to check that the transfer of the proteins was successful. All incubation steps described are carried out with gentle shaking.

2 The nitrocellulose filter is washed in $1 \times$ PBS at room temperature for 10 min.

3 The filter is then incubated in blocking solution (3% gelatine in $1 \times$ TPBS or 5% BLOTTO in $1 \times$ TPBS) for 1.5–2 h at 37 °C. This step reduces the unspecific binding of the antiserum to the filter. 3% BSA can also be used as blocking solution.

4 The filter is washed twice for 5 min in $1 \times$ TPBS at room temperature.

5 The filter is then sealed in a plastic bag containing 20 ml antiserum buffer (1% gelatine / $1 \times$ TPBS plus a pinch of NaN$_3$, or 0.5% BSA / $1 \times$ TPBS 0.01% (v/v) Thiamersol) and 50 μl antiserum. The bag is drawn over the edge of the lab bench to remove any air bubbles before sealing. The incubation with antiserum is carried out for at least 30 min but may also take place overnight at 37 °C.

6 The filter is then washed twice for 5 min in $1 \times$ TPBS to remove any unbound antiserum.

7 The filter is incubated for 2 h in 50 ml antiserum buffer containing 12 μl protein A / horseradish peroxidase conjugate.

8 The following washes are then carried out: twice 5 min in $1 \times$ TPBS, 5 min in $1 \times$ PBS and 5 min in 50 mM K phosphate buffer, pH 7.5.

9 Working in the dark, the filter is incubated for 5 min in freshly prepared diaminobenzidine solution containing 1.5 ml 1% CoCl$_2$ and 1.5 ml 1% Ni(NH$_4$)$_2$(SO$_4$)$_2$.

10 The staining reaction is started by the addition of 8 μl H$_2$O$_2$ and the incubation is continued until a blue colour is observed (this generally takes 5 min).

11 The filter is then washed twice in K phosphate buffer, pH 7.5, and once in dH$_2$O, placed on Whatman 3MM paper to dry and then photographed (Polaroid MP4 camera, aperture 8, exposure 1/30 s). The filter can be stored in the dark.

Indirect Method

The method is described in the booklet accompanying Promega's Protoblot (Western blot HRP system):

1   Details as above; the filter is washed in 1 × TBST.

2   The filter is incubated in blocking solution (1% BSA in 1 × TBST).

3   Incubation in antiserum for 30 min (1 × TBST + diluted antiserum).

4   The filter is washed three times, 10 min each, in 1 × TBST.

5   30 min incubation in anti-IgG / alkaline phosphatase conjugate in 1 × TBST containing a 1 : 7500 dilution of the second antiserum.

6   The filter is washed three times for 10 min each, in 1 × TBST.

7   The staining reaction is carried out by incubating the filter in 50 ml alkaline phosphatase buffer containing 330 μl NBT and 170 μl BCIP for 30 min.

8   The filter is washed in dH$_2$O, dried and photographed as described above.

Further reading

Western blot AP (alkaline phosphatase) and Western blot HRP (horseradish peroxidase) systems (Promega technical manuals).

Western blotting technical manual. (Amersham).

BITTNER, M., KUPFERER, P. and MORRIS, C.F. (1980) Electrophoretic transfer of proteins and nucleic acids from slab gels to diazobenzyloxymethyl cellulose or nitrocellulose sheets. *Anal. Biochem.* **102**, 459–71.

HARLOW, E. and LANE, D. (1988) *Antibodies: a laboratory manual* (Cold Spring Harbor, NY: Cold Spring Harbor Laboratory Press).

JOHNSON, D.A., GAUTSCH, J.W., SPORTSMAN, J.R. and ELDER, J.H. (1984) Improved technique utilizing non-fat dry milk for analysis of proteins and nucleic acids transferred to nitrocellulose. *Gene Analysis Techniques* **1**, 3–8.

POLLARD-KNIGHT, D. (1990) Current methods in non-radioactive nucleic acid labelling and detection technique. *Cell Molec. Biol.* **2**, 113–32.

ROSWELL, D.F. and WHITE, E.H. (1978) The chemiluminescence of luminol and related hydrazides. *Methods Enzymol.* **57**, 409–23.

TOWBIN, H., STAEHLIN, T. and GORDON, J. (1979) Electrophoretic transfer of proteins from polyacrylamide gels to nitrocellulose sheets: Procedure and some applications. *Proc. Natl. Acad. Sci. USA* **76**, 4350–4.

5.4   ANIMAL TISSUE CULTURE

M. Hülscher

General Comments

Animal tissue culture is an established technique in biotechnology and, as such, falls within the scope of this book. It is, however, virtually impossible to introduce examples of every method in the space available: several books have been written on the subject of animal tissue culture in the laboratory. The majority of tissue culture techniques are quite different from those of microbiology, and this is reflected in the scaling-up process. The following is a list of requirements for setting up a tissue culture facility adequately equipped for carrying out the experiments described below.

- The tissue culture facility should be quite separate from the area used for microbiological experiments; desirable, though not absolutely essential are separate areas for media preparation and sterilisation. It goes without saying that dedicated equipment for incubation as well as for harvesting the cells and the isolation of the desired product is essential. The reason for these precautions is the high risk of microbial contamination.

- The working area should be planned bearing in mind the strict requirement for sterility; i.e. the incubators should be as close as possible to the bench working area and in close proximity to the 'clean bench' and the fermenter.

- The basic requirements for tissue culture (cultured cell lines, CO_2-incubators, inverse microscope, etc.) will be assumed and not referred to further.

- A sterility control system must be set up; this involves quarantine for media and regular checking of the cultures for their identity and any signs of contamination.

- Good microbiological practice should be taught and observed when working with animal tissue culture.

It is appropriate at this stage to recommend the book: *Culture of animal cells* by R.I. Freshney (1987).

5.4.1 Preparation of Monoclonal Antibodies from Hybridoma Cells Grown in an Airlift Fermenter

Aim of the Experiment

The purpose of this experiment is to grow hybridoma cells for the production of monoclonal antibodies. Applications for monoclonal antibodies include their use as high-specificity reagents for analytical purposes, in affinity chromatography and in biomedical research. The experiment has a twofold purpose: on the one hand it describes the production of monoclonal antibodies for use in the laboratory and on the other hand it serves as an introduction to the methods and requirements of large-scale tissue culture. A secondary (and intentional) aim is the insight gained into the complexity of the metabolism involved in animal tissue culture.

Experimental Set-up, Reactors and Cell Lines

An essential piece of equipment for this experiment is a suitable reactor vessel. Airlift reactors have proved useful in many industrial production systems and for this reason this type has been chosen for the experiment described here. In principle other types of reactor can be used: e.g. baffle fermenter with filter aeration. If an airlift fermenter is used, certain requirements must be fulfilled: it should have a capacity of at least 6 l and the ratio height:diameter should lie between 5 and 6. The medium in the vessel should reach a height of at least 70 cm. Ideally, the ratio of the diameter of the inner feeder tube to that of the reactor should be 0.7. Aeration is best provided by nozzles which disperse air bubbles rather like a water spray pump directly into the feeder tube. The following dimensions and set-up are based on using a reactor with a capacity of 6 l. The nozzles should have seven jets each with a diameter of 0.2 mm and three if their diameter is 0.5 mm. The latter configuration is less prone to contamination with dirt particles and therefore appropriate for use in practicals. The reactor should be fitted with a valve allowing the addition of at least CO_2 to the air supply and, if possible, N_2 and O_2. The valve should allow the composition of the gas flow to be altered without altering the volume. The pH is controlled by means of a pH probe which monitors the CO_2 content of the aeration mixture. An oxygen electrode measures the partial O_2 pressure in the fermentation culture and pO_2 is regulated via the partial O_2 pressure in the gas phase. The recommended rate of aeration is 0.04 vvm. The modular construction of reactors favoured in microbiological biotechnology is not to be recommended for cell culture, on account of the high risk of contamination. Sterile connections should, whenever possible, be of a single piece. Fig. 5.4 illustrates the recommended version of a sterile connection. Closed systems should be used whenever samples are taken; Fig. 5.5 illustrates such a system. If it is not possible to install such a system the sampling tube should be placed under a 'clean bench', and that part of the tubing which is unsterile should be immersed in a disinfectant solution (e.g. 70% ethanol) between withdrawal of samples. Furthermore, a sterile connection for filling the reactor with filter-sterilised medium and a vessel for the short-term storage of samples withdrawn from the reactor should also be available. Fig. 5.6 shows a schematic representation of the reactor set-up. Reactors of this type are available from several companies including Braun, Melsungen.

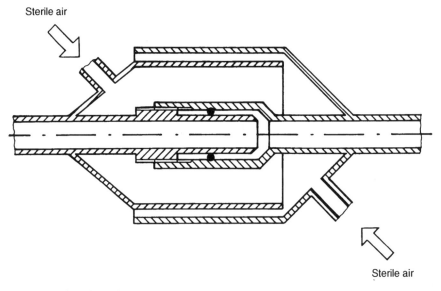

Sterile air

Sterile air

Figure 5.4 Enclosed sterile connection for tissue culture. The actual connection is screwed together and sealed with an O-ring. Outer coverings are placed over both ends of the tubing to ensure that the connection is contained in a sterile atmosphere until the end of the experiment, thus reducing the chance of contamination. When constructing such a connection, it must be remembered that the outer covering should be free to move around the central axis of the tubing.

From fermenter

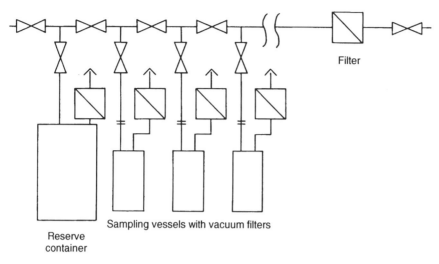

Filter

Sampling vessels with vacuum filters

Reserve
container

Figure 5.5 Closed sample-withdrawal system for tissue culture systems; presample and sample are withdrawn aseptically via membrane valves. The connecting tubes can be evacuated via the filter and valve shown in the top right of the illustration and flushed with compressed air.

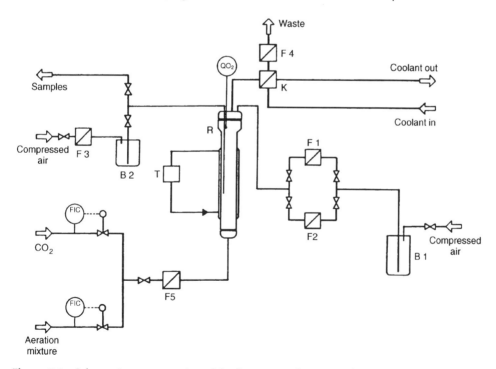

Figure 5.6 Schematic representation of the fermenter. The minimal requirements are shown in the diagram; measurement of pH, and its regulation via the CO_2 content of the airflow, as well as a gas mixing station, as described in the text, are to be recommended. For the sake of clarity, those parts of the fermenter necessary for steam sterilisation have not been included in the diagram. B1, container for medium; B2, container for surplus medium; F1–5, membrane filters; K, cooling system; R, airlift loop reactor; T, thermostate.

For the preparation of the preculture, CO_2 incubators and 'spinner' flasks with a capacity up to 2 l are essential. 'Spinner' flasks stirred by a rod which has been shaped to a ball at one end are to be recommended and are available from Techne. The first preculture can be set up in standard plastic tissue culture flasks as manufactured by Becton-Dickenson.

Mouse × mouse hybridoma cell lines are most frequently used in the production of monoclonal antibodies. In principle it is possible to use any cell line for this experiment; however, the fermentation conditions must be adjusted accordingly. The cell line W6/32 (ECACC No. 84112003) is recognised as a cell line suitable for use in practicals and is available from the European Collection of Animal Cell Culture (PHLS Centre for Applied Microbiology and Research, Vaccine Research and Production Laboratory, Porton Down, Salisbury, Wiltshire, SP4 0JG, UK).

Analytical Measurements

Cell Concentration

Equipment: Cell counting chamber, reaction tubes, Pasteur pipettes.

The number of cells is determined by counting a small aliquot of the culture in the counting chamber. The cells are used at a concentration for which a dilution is unnecessary.

Concentration of Live and Dead Cells

Equipment: as above and an Eppendorf pipette adjustable from 20 to 100 μl.
Solutions: *PBS* (8 g/l NaCl, Na_2HPO_4 1.15 g/l in dH_2O); *Trypan blue* (400 mg Trypan blue dissolved in 100 ml PBS).

100 μl cultured cells and 25 μl Trypan blue solution are mixed carefully and left to stand for 1 min. The sample is mixed again and one drop is placed in the counting chamber. Following this treatment, the dead cells appear blue and the living cells do not take on any colour.

Antibody Concentration

Material required: plate reader for ELISA tests equipped with filters for 492 and 600 nm; flexible 96-well microtitre plates (Becton-Dickenson); 8-channel Eppendorf pipette adjustable from 50 to 200 μl; Eppendorf pipettes covering the range 1–1000 μl; several wash bottles; plastic petri dishes.

Special reagents: commercially available rabbit antisera against mouse immunoglobulin, mouse immunoglobulin conjugated with peroxidase, and polyclonal mouse immunoglobulin (all available from Sigma).

Solutions required: *washing buffer* (1.5 g Tween 20, 3 l PBS; the boiled solution can be kept in the cold for several days); *blocking buffer* (5 g bovine serum albumin in 500 ml washing buffer; the solution must be made up freshly each day); *overlay solution* (4.2 g $NaHCO_3$ in 500 ml ddH_2O; stable for several weeks when stored in the cold); *sodium citrate buffer* (18.9 g $Na_3C_6H_5O_7 \times 5.5\ H_2O$ dissolved in ddH_2O; adjust the pH to 5 by the addition of concentrated citric acid; filter sterilise and store refrigerated); *o-phenylenediamine solution* (160 mg o-phenylenediamine dissolved in 10 ml ddH_2O, dispense 1 ml aliquots into reaction tubes and store, protected from light, at $-20\,°C$; *developer solution* (1 ml *o*-phenylenediamine solution, 9 ml sodium citrate buffer and 5 μl 30% H_2O_2 solution; to be made up immediately prior to use); *4.5 N sulphuric acid*.

Measurement range: 2–20 pg/μl; standards: polyclonal mouse immunoglobulin (Sigma); six standards covering the measurement range. Standard solutions and sample dilutions should be made in blocking buffer.

Hint. Dispense the washing buffer into a wash bottle and use to fill the wells of the micro-titre plate for the washing steps, and empty the plate into a large collecting vessel. To fill the wells with identical solutions (i.e. all steps except for pipetting the sample) the solutions can be decanted into plastic petri plates and transferred to the micro-titre plate with the 8-channel pipette. It goes without saying that a separate plate should be used for each solution.

Experimental Method

1   Add 10 μl undiluted anti-mouse-immunoglobulin rabbit antiserum to 10 ml wash solution, mix, and pipette 100 μl into each well of the micro-titre plate. Incubate either overnight at 4 °C or 1 h at 37 °C.

2   Wash three times with washing buffer.

3   Pipette 200 μl blocking buffer into each well of the micro-titre plate and incubate for 1 h at 37 °C.

4   Wash once with washing buffer.

5   Dilute samples with blocking buffer so that the final concentration lies within the measurement range. Pipette 200 μl of the diluted samples and the standard solutions into the wells of the micro-titre plate. Incubate at 37 °C for 2 h.

6   Wash three times with washing buffer. It is important at this step to ensure that there is no running of the solution from one well into another.

7   Pipette 10 μl peroxidase-conjugated anti-mouse-immunoglobulin rabbit antiserum into 10 ml washing buffer and mix thoroughly. Dispense 100 μl into each well of the micro-titre plate. Incubate for 1 h at 37 °C.

8   Wash three times with washing buffer and three times with ddH_2O.

9   Prepare developer solution and dispense 100 μl into each well of the micro-titre plate. Incubate for 15–30 min at 37 °C until the standards can be differentiated easily from one another. Stop the reaction by the addition of 25 μl 4.5 N sulphuric acid per well.

10   Determine the difference in extinction at 492 and 600 nm by means of the plate reader.

11   Determine the antibody concentration of each sample with the aid of the calibration curve obtained with the standards.

Glucose, Lactate and Glutamine Concentration

These substrates and products can be measured with commercially available tests. The following are recommended: glucose (Sigma No. 510-DA), lactate (Boehringer Mannheim No. 139 084), glutamine (Boehringer Mannheim No. 139 092) and asparaginase (Boehringer Mannheim No. 102 903).

MYCOPLASMA TEST

Samples taken at the beginning, during and at the end of the fermentation should be tested for contamination with mycoplasma. A relatively easy to use test, Mycotect, is available from Life Technologies. The basis of the test is the adenosine phosphorylase activity present in the medium.

Procedure

PREPARATION OF MEDIA

If the recommended cell line W6/32 is used the medium described below should be made up. This medium has been used successfully with other mouse × mouse hybridoma lines but can in no way be regarded as a universal medium. If in doubt, other well-established media should be chosen for other cell lines; however, it should be remembered that media containing more than 5% serum can be difficult to work with in an airlift reactor because of a tendency to foam building.

BASIC MEDIUM

A 1:1 mixture of IMDM from Life Technologies and Ham's F12 from Seromed/ Biochrom. Additives (per litre medium): glucose (Merck) 1350 mg, sodium hydrogen carbonate (Merck) 2000 mg, bovine serum albumin (Cohn fraction V, Sigma) 1000 ng, human transferrin (Sigma) 10 mg, bovine insulin (Sigma) 10 mg, ethanolamine (Sigma) 1.2 mg, sodium selenite (Sigma) 3.5 mg, ferric chloride × 6 H_2O 51 μg. The last three named substances should be made up as a high-concentration stock solution.

The medium should be stirred until no particles are visible to the naked eye. The medium is now filter sterilised (using a filter unit from Sartorius, pore size 0.2 μm) directly into the reserve medium chamber. The medium must be stored at 4 °C while sterility control tests (incubation of samples at 37 °C over a period of at least 4 days) are carried out. The medium should be used only after this test has been performed. Before filling the fermenter, the aeration must be switched on.

PRECULTURES

These should be started with 5 ml cultures (in 25 ml flasks) which ideally have been innoculated from the stock culture of the chosen cell line. When this culture has reached the late log phase (ca. $5-8 \times 10^5$ cells/ml) it is used to inoculate three 5 ml cultures at a cell density of 1×10^5/ml. After two days incubation (37 °C, 5% CO_2, closed atmosphere) a cell density of $5-8 \times 10^5$ has been reached again. These cells are the starting material for a 75 ml culture (in a 225 ml flask) which two days later serves as the inoculum for three 75 ml cultures. These cultures are the starting material for the 1 l 'spinner' culture.

All in all, a total of 8 days is required for the precultures. The 'spinner' culture is the inoculum for the fermenter.

PREPARATION OF THE FERMENTER

The fermenter should be prepared at the same time as the 'spinner' culture is inoculated. The preparation involves the setting-up and sterilisation of the empty fermenter before filling it with sterile medium. The medium should be filtered directly into the sterilised fermenter, and the aeration (5% CO_2 at the entry point) must be switched on. The fermenter must be connected to a sterile reserve container of at least 2 1 capacity into which medium from the reactor can be pumped and stored under sterile conditions. This can be achieved as follows: the fermenter is filled to the recommended volume, i.e. so that the circulation of the medium is guaranteed. Immediately before inoculation of the fermenter 1 1 of medium is transferred to the reserve container and the 1 1 inoculum 'spinner' culture is added to the medium in the fermenter. During the fermentation the volume in the fermenter can be maintained at a constant level by pumping medium from the reserve container (see Fig. 5.6, container B2).

Before inoculating the fermenter it is essential to calibrate the O_2 and pH electrodes. The O_2 electrode is checked by aerating the fermenter with a mixture of nitrogen and air and the pH electrode by varying the partial pressure of CO_2 and then checking pO_2 and pH in a sample of medium.

INOCULATION OF THE FERMENTER

When the 'spinner' culture has reached a cell concentration of about 8×10^5 cells/ml the fermenter should be inoculated according to the instructions given. The first sample is taken as soon as the working volume of the fermenter has been adjusted. The actual sample should have a volume of 10 ml, and the presample volume ought to be about 15 ml.

TREATMENT OF THE SAMPLE

The sample (10 ml) is transferred aseptically to a sterile centrifuge tube. Approximately 200 μl is removed and used to determine the number of cells/ml, and a second aliquot of 200 μl is treated with Trypan blue. The remainder of the sample is centifuged for 5 min at 1000 rpm. At least 5×1 ml aliquots of the supernatant are stored at $-20\,°C$ to be used for the analysis of antibodies, glucose, lactate and glutamine. The pH of the sample is determined and a test for contamination with mycoplasma is carried out. Samples should be taken at 6-hourly intervals. The values obtained from each sample measurement are to be presented in the form of a graph. The pH of the fermenter culture should be 7.4 and the CO_2 content of the air supply should be measured as a function of time. If the percentage of dissolved O_2 falls below 30% of the saturation value and there is no information to the contrary the amount of added O_2 should be increased. A reduction in the amount of dissolved O_2 should not occur when the cell line W6/32 is used, and when it does happen it is an indication that the culture is contaminated. Any increase in the air flow should be avoided.

TIME COURSE OF THE FERMENTATION

If the cell line W6/32 has been used the maximum cell concentration is attained after about 50 h, at which time the stationary phase of the culture begins. Roughly 3–5 h before the start of the stationary phase the amount of dissolved O_2 reaches a minimum before it commences a rapid increase. The experiment can be stopped after 120 h and the cells harvested. The supernatant can be used directly for analytical purposes; the methods to be used for the isolation of antibodies can be found in the literature. It must be emphasised again that the use of a cell line other than W6/32 may lead to different values for the given parameters.

All waste must be autoclaved. Small volumes can be overlaid with 70% ethanol if they have to be stored before autoclaving. Once it has been emptied the fermenter must be sterilised by autoclaving or thoroughly rinsed out by spraying with ethanol.

Environmental and Safety Aspects

Any one working with tissue culture must be familiar with good microbiological practice and follow the safety rules applicable for microbiological laboratories. All students must be informed of, and trained in, these safety regulations. There is a very limited but nevertheless potential risk of viral infection from contaminated cultures and additives. In the experiment described here the danger lies in the cell line used, the protein components of the media and, following filter-sterilisation, the medium itself. Transferrin which is isolated from human blood samples should be handled with the utmost care. Appropriate safety measures include wearing gloves and dedicated lab coats as well as the chemical disinfection of the lab surfaces and the aforementioned high-temperature sterilisation of waste and any equipment coming into direct contact with the culture. Experienced members of the laboratory have the responsibility of supervising and advising those less experienced or new to the field.

The experiment requires relatively large amounts of plastic ware (culture vessels, micro-titre plates, etc.) and from the point of view of protecting the environment the amount used should be kept to a minimum. Multiple usage should be practised whenever possible. On account of the effects of surface aging, the use of glass products is not to be recommended.

Presentation and Interpretation of the Results

The values obtained for the cell concentration must be presented as a semi-log plot over the time course of the experiment. The concentrations of substrate and product are analysed at the end of the fermentation. These results, together with those obtained for pH and dissolved O_2, should be plotted against time. The results can be presented pairwise, e.g. glucose and lactate concentrations, glutamine concentration and dissolved O_2 quotient and pH, CO_2 content of the air flow and lactate concentration, in the same diagram. The antibody concentration should be presented together with the number of viable cells counted. For each sample the quotient lactate produced / glucose used should be calculated. In the discussion particular attention should be paid to indications that suggest a connection between any two parameters. Hints for such correlations are the suggestions given above for pairwise presentation of the results.

References

European Collection of Animal Cell Cultures, *Catalogue of Cells and Services*, 4th edn.
FRESHNEY, R.I. (1987) *Culture of animal cells*, 2nd edn (New York: Alan R. Liss).

HÜLSCHER, M. (1990) *Auslegung von Airlift-Schlaufenreaktoren für die Kultivierung von Tierzellen in Hinblick auf Zellschädigungen.* VDI-Fortschritt-Berichte, Reihe 3, Nr. 229 (Düsseldorf: VDI-Verlag).

JOHANNSEN, R., ALBERT, W., HUNSMANN, G., KRÄMER, P., NOÉ, W., SCHIRRMACHER, V., SCHLUMBERGER, H.-D. and STREISSLE, G. (1988) Chancen und Risiken durch Säugerzellkulturen. *Forum Mikrobiologie* **11**, 359–67.

5.4.2 Culturing Chicken Embryo Fibroblasts in Microcarrier Culture

Aim of the Experiment

The majority of animal cells when propagated in culture require a surface to which they can adhere. A common and easy-to-scale-up method of providing such surfaces is the microcarrier culture. This method will be explained and put into practice as described below. The experimental procedure may have to be altered in some respects if a different cell line is used but the principle of the method is always the same. The procedure described here is that for chicken embryo fibroblasts – a cell type which is easily obtained and is employed, albeit to a limited extent, in the industrial production of vaccines. The aim of the experiment is to gain familiarity with the techniques involved in microcarrier culture.

Experimental Set-up, Cells and Medium

The instructions given are for a 1 l culture. 'Spinner' cultures will be used, although a 1 l fermenter culture could be handled in the same way. For economic reasons 'spinner' cultures are recommended for practical courses.

The 'spinner' flasks (Techne) should have a volume of 2 l. The stirrer is a magnetically driven stab formed to a bulb at its lower end. A low-speed magnetic stirrer is used to drive the stirrer. Stirrer and flask are placed in a CO_2 incubator. A supply of sterilised glass containers of varying sizes should be available.

The cells can be prepared from chicken embryos or obtained commercially (see previous section). Instructions for the preparation of the cells from chicken embryos are beyond the scope of this chapter, and it is therefore recommended to consult the appropriate section in Freshney (1987) and information available from Pharmacia. If the cells are obtained commercially (to be recommended for practical courses) the supplier's instructions regarding medium and handling should be followed. The recommended medium is DMEM (FLOW-ICN) enriched with 1% chicken serum, 5% foetal calf serum and 10% tryptose phosphate broth. The sera should be purchased from Life Technologies or Sigma.

Methods for Checking the State of the Culture

Direct Microscopic Observation and Photographic Documentation

When working with microcarrier cultures, direct microscopic observation coupled with photographic documentation is an important method for checking the status of the culture. The culture sample (the carrier must be well suspended) is placed in the wells of a 6-, 12- or 24-well micro-titre plate or on a microscope slide. In the latter case the coverslip should be broken into fragments before placing on the culture.

Cell Counting

Equipment: counting chamber, pasteur pipettes, adjustable Eppendorf pipette 100–1000 μl, culture plates with a volume of 5 ml, centrifuge tubes (10 ml).

Reagents: Ca^{2+}- *and* Mg^{2+}-*free PBS* (8 g sodium chloride, 1.15 g disodium hydrogen phosphate dissolved in 1000 ml MilliQ water; adjust pH to 7.6 before autoclaving); *EDTA* (0.2 g EDTA, Ca^{2+}- and Mg^{2+}-free PBS ad 1000 ml); *trypsin solution* (0.5 g trypsin, Ca^{2+}- and Mg^{2+}-free PBS ad 200 ml).

1 ml aliquot of resuspended microcarrier culture is placed in a culture plate. The microcarrier is allowed to precipitate and the 'supernatant' is removed carefully with a Pasteur pipette and discarded. 2 ml EDTA is added to the microcarrier and the whole is resuspended. The microcarrier is allowed to precipitate and the 'supernatant' is again removed with a Pasteur pipette and discarded. 0.5 ml EDTA and 0.5 ml trypsin solution are added to the microcarrier. Place at 37 °C for 15 min; during the incubation the suspension should be re-suspended gently from time to time. The microcarrier is allowed to precipitate and the 'supernatant' is removed by aspiration and placed in a centrifuge tube. The microcarrier is washed twice with 1 ml culture medium, allowing the microcarrier to sediment each time and transferring the 'supernatants' to the centrifuge tube.

The cell suspension is centrifuged at 1000 rpm for 5 min and the pellet is resuspended in 2 ml Ca^{2+}- and Mg^{2+}-free PBS and counted immediately in the counting chamber. The total cell concentration is obtained. REMEMBER THE DILUTION FACTOR OF 2!

Proportion of Living and Dead Cells

Equipment: as above with an additional adjustable Eppendorf pipette (20–100 μl) and reaction tubes.

Solutions: *PBS* (8 g/l NaCl, 1.15 g/l Na_2HPO_4, dissolved in MilliQ water); *Trypan blue solution* (400 mg in 100 ml PBS).

Procedure: 25 μl Trypan blue solution is added to 100 μl of the cell suspension obtained as described in the previous subsection and the whole is mixed gently. A small aliquot is placed in the counting chamber. The dead cells appear blue and the living cells remain colourless. After prolonged contact of the cells with Trypan blue the dye can be taken up by living cells.

Procedure

TREATMENT OF THE CULTURE FLASKS

The culture flasks must be siliconised to prevent the microcarrier from adhering to the sides of the vessel. Stainless steel fermenter vessels do not require siliconising. Solutions of dimethyldichlorosilane in organic solvents are used for siliconising; the product *Sigmacote* (Sigma) has proved to be reliable. The completely dry vessel is rinsed out with the smallest amount possible of the solution so that all the surfaces are coated. The liquid is allowed to drain out and the vessel left to dry before autoclaving. The siliconised vessels can be used several times before the process has to be repeated. Pipettes and other equipment coming into contact with medium should also be siliconised.

PREPARATION OF THE MICROCARRIER

3 g Cytodex-1-microcarrier (Pharmacia) is soaked in 300 ml Ca^{2+}-and Mg^{2+}-free PBS for 3 h. The contents of the container should be swirled around from time to time. The PBS is poured off and the microcarrier is washed twice with 150 ml Ca^{2+}- and Mg^{2+}-free PBS. 150 ml Ca^{2+} and Mg^{2+}-free PBS is added and the whole is sterilised for 15 min at 120 °C and 1 atu. After autoclaving, the liquid is removed carefully with a pipette and the microcarrier is washed once with 100 ml sterile medium. The microcarrier is then resuspended in 300 ml medium and dispensed (large-volume pipettes should be used) into the culture vessels.

CULTURING OF CELLS

2×10^8 cells (prepared either from chick embryo or obtained commercially) are added to the prepared microcarrier suspension. Incubation is carried out at 37 °C. During the first 3–6 h of the incubation the culture should be stirred at 10–20 rpm and thereafter at 50 rpm. Alternate phases of stirring and resting are also recommended in the first 3 h of the culture if the magnetic stirrer has the appropriate function. Samples should be taken every 6 h to determine cell concentration (live, dead and total), pH and to obtain a microscopic picture of the culture. After 24 h incubation 200 ml fresh medium is added and again after a further 24 h. On the third day 300 ml fresh medium is added. On the fourth day 500 ml of medium is removed from the vessel, taking the utmost care to avoid removing any of the microcarrier, before adding 500 ml fresh medium. The culture should be confluent by the eighth day and have a concentration of about $1–3 \times 10^6$ cells/ml. The experiment is now finished.

WASTE DISPOSAL

For reasons of safety, all waste must be autoclaved. The remains of the samples should be diluted with an excess volume of 70% ethanol and autoclaved within two days. The empty culture vessel should be autoclaved.

Environmental and Safety Aspects

The comments made in Subsection 5.4.1 also apply to this experiment. A possible source of danger is the use of sera of inferior quality. It cannot be emphasised enough that sera should be obtained only from reliable sources. All procedures must be carried out on a clean bench. The amount of plastic ware required is reasonable.

Interpretation

Since the purpose of the experiment is to learn the handling techniques, the interpretation of the experiment is restricted to the presentation of the data obtained. A representation of the cell concentration is not to be recommended. More information is obtained with a representation of the number of cells in the culture vessel and the cell concentration (live and dead) on the microcarrier in cells/cm^2 with respect to time, pH and the volumes of medium added and removed throughout the period of cultivation. The microscopic picture should be interpreted with respect to the growth curve. Tests for mycoplasma should be carried out.

Further reading

Microcarrier cell culture – principles and methods (Pharmacia).
FRESHNEY, R.I. (1987) *Culture of animal cells*, 2nd edn (New York: Alan R. Liss).
JAKOBY, W.B. and PASTAN, I.H. (1979) *Cell culture Methods Enzymol.*, **58**: *Cell culture* (New York: Academic Press).

5.5 PLANT CELL AND TISSUE CULTURE

A.W. Alfermann, K. Dombrowski and M. Petersen

In this section several experiments will be described which illustrate the biotechnological application of plant cell culture. Space permits that only those applications of cell culture used in the production and the biotransformation of natural products can be discussed. Cell culture also has considerable economic importance in the propagation of plants, for the production of pathogen-free plant material and plant breeding in general.

Information on possible experiments for these topics can be found in the books written by Dixon (1985), Dodds and Roberts (1985) and Seitz *et al.* (1985); summaries are to be found in Bhojwani and Razdan (1983), Fowler and Warren (1992), George and Sherrington (1984) and in the multi-volume works on plant cell and tissue culture by Bajaj (1985), Bonga and Durzan (1987), Evans *et al.* (1983) and Vasil (1984).

5.5.1 Setting-up and Subculturing of Callus and Cell Culture

Culture Media

Specific culture media have been developed for many types of plants. Many are based on the media developed by Murashige and Skoog (1962) for tobacco callus cultures or by Gamborg *et al.* (1968) for soya. The media used (see Table 5.1) in the experiments described here are also based on these original recipes. Other variations are to be found in the books on cell culture listed above. A collection of culture media for a large number of plants is to be found in the books by George *et al.* (1987 and 1988). In the composition of the media one generally differentiates between inorganic macro- and micro-elements and additives such as vitamins, plant growth factors, carbon source and additional amino acids. The amino acids can be added in defined amounts as the pure substance or in the form of an acid hydrolysate of milk protein.

Setting-up a Callus Culture

The setting-up of a callus culture is the basic experiment in which the cultures required for the experiments described here are established. It is possible to establish callus cultures from many types of plants and from different parts of the plant. A piece of plant tissue is removed aseptically and incubated in nutrient medium which contains, in addition to inorganic salts, a carbon source (usually sucrose), several vitamins, inositol, amino acids and, most importantly, growth hormones belonging to the classes auxins and cytokines, at concentrations derived experimentally. For the purposes of example, the setting-up of two callus cultures – one from the carrot (*Daucus carota*, cv. Rotin) and the other from *Coleus blumei* will be described. Although it is possible to establish callus cultures from virtually any plant organ, the best results are obtained when the starting material is a sterile seedling. This procedure will be described for both types of plant before describing the setting up of callus cultures from cuttings and leaves of *Coleus blumei*. All steps involved in transfer and setting up the callus cultures should be carried out on a 'clean bench'. This makes it easier to work aseptically. It is, however, also possible to perform the experiment in a laboratory which is essentially sterile and using an 'inoculation box'.

Establishing Sterile Seedlings of *Daucus* and *Coleus*

Equipment Required for Each Batch of Sterilisation

Six sterile glass or plastic petri dishes, about 10 cm diam.; five of these plates are placed on the 'clean bench' and each filled to a depth of about 10 mm as described below.

- 70% ethanol is poured into one plate.
- A second plate should contain a solution of sodium hypochlorite (commercially available hypochlorite solution with 12% bleach is diluted 1 : 10) or a solution of Dimanin C (5 g/l). Dimanin C (Bayer) is used for chlorinating swimming pools and, even in dilute solution, should be treated with the utmost care.
- Three plates are filled with distilled water.
- The sixth plate is used to store the sterilised seeds. Instead of petri plates it is possible to use 1 oz screw-top bottles. These bottles are easier to handle than petri plates but care must be taken that the tops can be autoclaved. If screw-topped bottles are used, the water can be sterilised in them. On account of the danger of implosion, the tops are not screwed down tightly before autoclaving. The sterile bottles are allowed to cool on the clean bench before the tops are screwed on firmly.
- Fine forceps are used for laying out the seeds. The forceps (several pairs) are wrapped singly in aluminium foil and autoclaved for 20 min at 121 °C or placed in a drying oven (180 °C) for 60 min. On no account should these fine forceps be placed in the flame of the Bunsen burner, since this causes the tips of the forceps to bend.
- Fine gauze is required in which to pack the seeds for sterilisation. Small nets can be made from nylon sieve material (gauge 0.1 mm).
- Twenty petri dishes (diam. 5 cm) are filled to a depth of 5 mm with hormone-free 12a or CB agar on the clean bench. Parafilm™ is required for sealing the petri dishes.

Procedure

- The seeds, packed in gauze or nylon, are placed in the petri dish containing 70% ethanol.

- The seed 'packets' are then transferred to the petri dish containing hypochlorite solution for 10–15 min. If it turns out that the seeds are still contaminated, the period of sterilisation should be increased. In other instances it may be necessary to reduce the period of sterilisation, to prevent excess damage to the seeds.

Table 5.1 Composition and preparation of CB and 12a media – variations of B5 and Murashige–Skoog media, respectively (Gamborg *et al.*, 1968; Murashige and Skoog, 1962). The concentrations given are the final concentrations in the media.

Substance	CB medium (mg/l)	12a medium (mg/l)
Macroelements[a]		
KNO_3	2500	1900
NH_4NO_3	–	1650
$MgSO_4 \times 7\ H_2O$	250	370
$NaH_2PO_4 \times H_2O$	172	–
$KH_2PO_4 \times H_2O$	–	170
$CaCl_2 \times 2\ H_2O$	150	440
$(NH_4)_2SO_4$	134	–
$FeSO_4 \times 7\ H_2O$	25.6[b]	27.8
Na_2-EDTA	34.3[b]	37.2
Trace elements[c]		
H_3BO_3	3	6.2
$ZnSO_4 \times 7\ H_2O$	3	8.6
$MnSO_4 \times H_2O$	1	16.9
KI	0.75	0.83
$Na_2MoO_4 \times 2\ H_2O$	0.25	0.25
$CuSO_4 \times 5\ H_2O$	0.25	0.25
$CoCl_2 \times 6\ H_2O$	0.25	0.25
Vitamins[d]		
Thiamine dichloride	10	0.1
Pyridoxine hydrochloride	1	0.1
Nicotinic acid	1	0.5
Hormones		
2,4-dichlorophenoxy acetic acid[e]	2	0.1
Naphthylacetic acid[e]	0.5	–
Indolyl-3-acetic acid[e]	0.5	2
Kinetin[f]	0.2	0.2
Other additives		
myo-inositol	0.1 g/l	0.1 g/l
NZ-amine[g]	2 g/l	amino acids[h]
Sucrose	20 g/l (CB_2) 40 g/l (CB_4)	30 g/l

All compounds are dissolved in ddH$_2$O. The media are made up to the final volume and the pH is adjusted to 5.5 with either 0.5 N HCl or KOH.

For solid medium agar–agar is added at a concentration of 10 g/l and the medium is heated to dissolve the agar completely. The medium is dispensed into 300 ml Erlenmeyer flasks: 50 ml aliquots for liquid media and 100 ml aliquots for solid medium. The flasks are closed with cotton wool plugs before autoclaving for 20 min at 117 °C.

Remarks

[a] The macroelements are made up singly in 100× stock solutions; 10 ml of the stock solution is added per litre of medium.

[b] Both these substances are made up as a single 100× stock solution which is heated to bring them into solution.

[c] The trace elements are made up as a 100× stock solution; 10 ml of the solution is added per litre of medium.

[d] The vitamins are made up as a single 1000× stock solution; 1 ml of this solution is added per litre of medium. The stock solution should be stored in 1 ml aliquots at −20 °C.

[e] These hormones are also made up as a 100× stock solution and stored frozen in 10 ml aliquots. To bring the hormones into solution they are dissolved in a small volume of absolute alcohol and the volume made up with ddH$_2$O. 10 ml of the stock solution is required per litre of medium.

[f] To prepare the 100× stock solution kinetin is dissolved in a small volume of 0.5 N HCl and then made up to the final volume with ddH$_2$O. The stock solution is stored frozen in the 10 ml aliquots required per litre of medium.

[g] NZ-amines are available from Sigma.

[h] For the preparation of 12a medium a 1000× stock solution of amino acids at defined concentrations is required. The final concentration of the amino acid in the medium is given below:

	mg/l
L-alanine	29.7
4-aminobutyric acid	26.0
L-arginine-HCl	3.1
L-asparagine	3.8
L-aspartic acid	1.7
L-cysteine	3.0
L-glutamic acid	15.7
L-glutamine	0.3
glycine	2.7
L-histidine	0.05
L-leucine	5.0
L-lysine-HCl	2.0
DL-methionine	0.05
DL-phenylalanine	0.05
L-proline	1.9
L-serine	12.8
DL-threonine	4.1
L-tyrosine	0.05
DL-valine	2.3

- From now onwards all steps must be carried out on the clean bench or in the inoculation box using sterilised forceps. The necks of the Erlenmeyer flasks must be flamed before removing the cotton wool plug. This is also important when preparing callus and suspension cultures and will not be referred to again!

- The hypochlorite solution is washed out of the seeds by placing the seed 'packets' sequentially in the three petri dishes containing sterile distilled water. The seeds are finally placed in the empty petri dish.

- 3–5 seeds are laid on hormone-free 12a medium (*Daucus*) or hormone-free CB medium (*Coleus*) in the petri dishes which are then sealed with Parafilm™ and incubated in the dark at 25 °C. For some species it is recommended that the seeds are placed in sterile water for a few hours before placing them on the agar.

- Several days later, when the majority of the seeds have germinated, the cultivation is carried out under illumination (1000–2000 Lux). Petri dishes contaminated with bacteria or fungi are discarded.

Preparation of Callus Cultures from Sterile Seedlings

EQUIPMENT REQUIRED

100 ml or 300 ml Erlenmeyer flasks containing 50 ml or 100 ml of the required culture medium, respectively, are sealed with cotton wool or gas-permeable silicone plugs or even with strong aluminium foil and autoclaved; 10–20 sterile petri plates (5 cm diam.), filled to a depth of 5 mm with sterile 12a or CB agar; scalpel and forceps as detailed in the previous section; long spatula.

PROCEDURE (TO BE CARRIED OUT UNDER STERILE CONDITIONS, SEE ABOVE)

- Using the scalpel and forceps, seed leaves or small first leaves, 5–10 mm strips of hypocotyl and epicotyl tissue, as well as pieces of root tissue are placed on 12a medium (*Daucus*) or CB medium (*Coleus*) in the petri dishes. The tissue explants can be 'pressed' onto the medium by light pressure of the forceps, the leaf and stalk explants can also be 'planted' into small slits in the agar.
- The explants are incubated at 25 °C either in the dark or under lumination (1000–2000 Lux). Calli develop, particularly at the cut surfaces, after several days to a few weeks.
- When the calli have a diameter of about 5 mm or, at the latest, after 8 weeks, they are transferred to fresh medium in Erlenmeyer flasks or culture bottles.
- From now on, a constant frequency (every 2–4 weeks) of re-inoculation should be maintained. The frequency of re-inoculation depends not only on the growth rate of the culture but also on the size of the inoculum. It is recommended that 1–2 g of callus is transferred to flasks containing 50–100 ml medium. The callus is broken up with the spatula and spread evenly over the surface of the agar.
- If the surface of the medium is overgrown or the callus turns brown, it has to be transferred immediately to fresh medium. Even if the culture is slow growing, the frequency of re-inoculation should not be too long otherwise the medium dries out.

Setting-up a Callus Culture from Leaf and Stem Pieces of *Coleus*

Preparation of Surface-sterilised Leaf and Stem Fragments

EQUIPMENT

Essentially the same equipment is required as for the preparation of sterile seedlings; gauze or nylon bags are not required but a scalpel is necessary. Furthermore, the glass bottles are more suitable for this experiment than petri dishes.

To reduce the risk of contamination several rounds of sterilisation should be performed, i.e. only a few leaves or stem fragments are sterilised in the same container.

PROCEDURE

- The leaves are removed from an approximately 10 cm long shoot. However, 1–2 cm long leaf stems should be left on the shoots.

- In addition, leaves about 8 cm long are removed or larger leaves are cut to size. Pieces of leaves should retain the midrib.

- Two or three leaves or pieces of stem are placed in 70% ethanol and then placed in the Diamin C solution. The time required for sterilisation depends on the extent to which the plants are contaminated and the sensitivity of the tissue used. Suggested sterilisation times are 5, 7.5, 10 and 15 min. If the explants lose their colour during the sterilisation or in the days immediately following the sterilisation then the sterilisation time must be reduced. If all explants are contaminated then the length of the sterilisation procedure must be increased.

- Finally the sterilised shoot and leaf pieces are washed three times with distilled water before placing them in a sterile container.

Setting-up the Callus Cultures

EQUIPMENT REQUIRED

The same equipment is required as for setting up callus cultures from seedlings of *Coleus*.

PROCEDURE

- 5 mm pieces of tissue are cut out of the pieces of shoot; the fragments from the ends of the shoots are discarded. Pieces of leaf tissue, about 5 mm^2, are removed from the central region of the leaf.

- The shoot slices and leaf fragments are pressed lightly onto the agar. They can be placed with the cut surface in contact with the agar, but this is not essential.

- Proceed as described in the experiment with seedlings.

Setting-up a Suspension Culture

EQUIPMENT REQUIRED

300 ml Erlenmeyer flasks containing 50 ml of the appropriate liquid medium (sealed as described above); inoculation spatula; sterile sawn-off glass pipettes (10 and 20 ml); sterile sieve (1 mm mesh) and a sterile glass or plastic funnel.

PROCEDURE

- Approximately 5 g of loosely packed callus tissue is transferred to liquid medium. If less than 5 g callus is available then the volume of medium should be reduced accordingly.

- The cells are incubated with or without illumination on a circular shaker with a 50 mm radius of movement at 120 rpm. If the radius of movement is less, the speed of shaking should be increased to 130–140 rpm.

- If the calli do not disintegrate properly into smaller cell aggregates they should be pressed through a tea strainer before the next transfer.

- After several passages it should be possible to transfer the cells with a sawn-off pipette as described in the following experiment.

- The transfers should take place every 7 days; if the cultures grow slowly then every 14 days is better.

5.5.2 Characterisation of a Suspension Culture of Coleus blumei Producing Rosmarinic Acid

Basic Principles

Coleus blumei belongs to the Lamiaceae family, in which the natural product rosmarinic acid, an ester of 3,4-dihydroxycinnamic acid and 3,4-dihydroxyphenyl-lactate is found. Rosmarinic acid is one of the active components of melisse, mint, sage and thyme. *Coleus* is a native plant of the tropical and sub-tropical regions of Asia and Africa, where it is used for medicinal purposes. Throughout the world there are numerous varieties of *Coleus blumei* which are used for decorative purposes on account of their colourful leaves.

Rosmarinic acid makes up between 2% and 3% of the dry weight in *Coleus*. When grown in cell culture in a high sucrose medium the plant can accumulate up to 20% of its dry weight as rosmarinic acid (Petersen, 1993). This is a culture with one of the highest known production rates for a secondary metabolite.

The aim of the experiment described here is to observe the growth, production of the secondary metabolite and any changes in the culture medium over a 14-day culture period of *Coleus blumei* in cell culture. At the same time the activity of the enzyme which connects primary cell metabolism with the synthesis of rosmarinic acid will be monitored; the enzyme phenylalanine ammonium lyase (PAL) oxidatively deaminates phenylalanine to *t*-cinnamic acid. The 3,4-dihydroxy-cinnamic acid moiety of rosmarinic acid is synthesised from phenylalanine by way of *t*-cinnamic acid, whereas the 3,4-dihydroxyphenyllactate moiety is derived from tyrosine. The suspensions of *Coleus* are cultivated in parallel in CB_2 (2% sucrose) and CB_4 (4% sucrose) medium to emphasise the effect of sucrose concentration on the production of rosmarinic acid.

During the course of the culture the following parameters will be monitored:

- wet weight
- dry weight
- cell number.

The activity of secondary metabolism is monitored by measuring

- rosmarinic acid content of the cells
- specific activity of the enzyme phenylalanine ammonium lyase.

This requires that the protein content of the enzyme suspension is determined.

The following parameters reflect changes in the culture medium:

- pH value
- sugar content as measured by the refractive index

- conductivity
- osmolarity
- phosphate content
- nitrate content.

Inoculation of the Culture

For the inoculation of the culture the following are required: clean bench; six 7-day-old cultures (70 ml each) of *Coleus blumei* in CB_2 medium; fifteen 300 ml Erlenmeyer flasks containing 50 ml CB_2 medium and fifteen containing CB_4 medium (see Table 5.1), sealed with cotton wool plugs and autoclaved for 20 min at 117 °C; cotton wool-plugged and sterile, sawn-off 10 ml pipettes; pipetting aid.

Using the sawn-off pipettes, 2×10 ml aliquots of the 7-day-old suspension cultures are transferred to the fresh medium in the Erlenmeyer flasks. The necks of the flasks must be flamed before and after removing the plugs. Prior to removing the aliquots the culture must be mixed thoroughly and the pipette should never come into contact with the bottom of the flask. If these precautions are taken, each flask will receive a homogeneous inoculum. The flasks are incubated on a rotary shaker (120 rpm) at 25 °C in the dark.

On each day of the 14-day incubation period, starting on the day of inoculation, one culture with CB_2 and one culture with CB_4 medium are used to measure the chosen parameters.

Determination of the Growth Parameters

The procedures for determining cell number, wet weight and dry weight (which may also be determined by lyophilisation) are to be found on pp. 12–14.

Secondary Metabolism (Determination of the Rosmarinic Acid Content of the Cells)

EQUIPMENT

1 ml and 2 ml glass pipettes; stoppered test tubes; 70% ethanol (v/v); heated ultrasonic water bath set at 70 °C; test tube rack; (the extraction can be carried out in a normal water bath at 70 °C if the tubes are shaken frequently and vigorously); table top centrifuge; pipette delivering 10 μl; disposable plastic cuvettes; spectral photometer.

The dry weight of the cells is determined (see p. 14) and 20 mg of freeze-dried cell material is weighed into the test tubes; 2 ml 70% ethanol is added to each test tube and the tubes are treated twice for 10 min at 70 °C in the ultrasonic water bath. It is essential that the cells are shaken vigorously throughout the procedure. To prevent evaporation the test tubes should be stoppered. The cells are then centrifuged for 10 min at 1000–3000 g. 10 μl of the ethanolic supernatant is placed in a cuvette together with 0.99 ml 70% ethanol, and the adsorption at 333 nm is determined. Rosmarinic acid has a characteristic adsorption at 333 nm. The reference cuvette contains 70% ethanol. The amount of rosmarinic acid can be calculated as follows:

$$RA\,(\%\,DW) = \frac{OD_{333} \times 379}{DW\,(\text{mg})}$$

One also requires the extinction coefficient of rosmarinic acid: 19 000 cm^2 mmol^{-1}; molecular weight of rosmarinic acid: 360.3 g/ml.

In addition to determining the rosmarinic acid content of the cells, it is possible to run spectra of pure rosmarinic acid and the ethanolic cell extract against 70% ethanol (Fig. 5.7).

Specific Activity of Phenylalanine Ammonium Lyase (PAL)

EQUIPMENT AND SOLUTIONS

Ice bucket; balance; spatula; plastic centrifuge tubes with a volume of approx. 30 ml; 2 ml glass pipettes; 0.2 M KH$_2$PO$_4$/K$_2$HPO$_4$ buffer, pH 8.0 containing 1 mM dithiothreitol (DTT) (DTT should be made up as a 1 M stock solution in ddH$_2$O and stored at -20°C). DTT is added to the amount of buffer required for 1 day at a dilution of 1 : 1000; Polyclar AT (insoluble polyvinylpyrrolidone, available from Serva); homogeniser (a pestle and mortar may also be used); refrigerated centrifuge; two small funnels plugged with a small amount of glass wool; two 10 ml test tubes; 0.1 M phenylalanine in 0.2 M potassium phosphate buffer, pH 8.0; four quartz cuvettes; spectral photometer with a thermostatically controlled cuvette holder (the work load is reduced considerably if an automatic cuvette changer and recording device are available for the photometer); Bradford solution for protein determination (see p. 10); protein standard (1 mg/ml bovine serum albumin in H$_2$O; can be stored at -20°C); disposable plastic cuvettes; Parafilm™; Eppendorf pipettes covering the range 10–1000 μl.

The preparation of the enzyme extract is carried out at 4°C. 2 g of cells are weighed into each centrifuge tube. 0.2 g Polyclar AT is added to each tube to remove phenolic compounds. 2 ml potassium phosphate buffer containing 1 mM DTT is added and the whole is homogenised in three 30 s bursts; between homogenisation treatments the tubes should be placed on ice for 30 s to prevent them from getting too warm. Alternatively the cells can be ground in a cooled morser placed on ice until a homogenate the consistency of pouring cream is obtained. The homogenate is then centrifuged at 4°C and at least 10 000 g for 15 min. The supernatant is filtered through glass wool into test tubes placed in ice. This enzyme extract is used to determine PAL activity and is assayed for its protein content. If these determinations are not to be carried out immediately, the extracts can be frozen at -18°C.

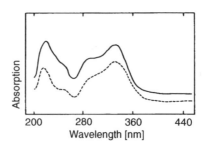

Figure 5.7 Spectrum of pure rosmarinic acid (—) and a 70% ethanolic extract (···) of freeze-dried cells from a suspension culture of *Coleus blumei* grown in CB$_4$ medium.

PAL activity is determined as the increase in the concentration of *t*-cinnamic acid produced from phenylalanine and can be followed photometrically at 290 nm. Two cuvettes are required for each enzyme extract; the reference cuvette and the sample cuvette differ only in that the former does not contain phenylalanine. The cuvettes contain:

- Sample cuvette:
  150 μl enzyme extract
  200 μl 0.1 M phenylalanine solution
  650 μl 0.2 M potassium phosphate buffer, pH 8.0
- Reference cuvette:
  150 μl enzyme extract
  850 μl 0.2 M potassium phosphate buffer, pH 8.0.

The cuvettes are 'closed' with Parafilm™ so that the contents can be mixed well by inversion. The cuvettes are then placed in the holder, which is kept at 36 °C. After a 10 min wait to allow the cuvette to reach a temperature of 36 °C the extinction is measured every minute over a period of 30 min. The values obtained are plotted against time and the increase in OD / 30 min is measured from the straight line drawn through the points. The enzyme activity is calculated according to the formula given below:

$$\text{PAL activity } (\mu\text{cat}) = \frac{ODV}{td\varepsilon_{290}}$$

where *OD* is the alteration in extinction over a period of 30 minutes, *V* is the volume in the cuvette ($=1$ ml), *t* is the reaction time ($=1800$ s), *d* is the light path of the cuvette ($=1$ cm), ε_{290} is the coefficient of extinction for cinnamic acid at 290 nm ($=10$ cm^2 μmol^{-1}).

In order to calculate the specific activity the protein concentration of the enzyme extract has to be determined. Protein is determined according to Bradford (see p. 10).

The specific activity of PAL is calculated as follows:

$$\text{Specific activity of PAL } (\mu\text{cat/kg}) = \frac{\text{PAL activity } (\mu\text{cat}) \times 10^6}{vP}$$

where *v* is the volume of enzyme extract used in the test system ($=0.15$ ml) and *P* is the protein concentration of the enzyme extract (mg/ml).

Determining Changes in the Culture Medium

The following parameters are measured in the culture supernatant remaining after the cells have been removed:

- pH (pH electrode)
- sugar content (see p. 8)
- conductivity (using a conductivity meter)
- osmolarity with the aid of an osmometer
- phosphate content (see p. 7)
- nitrate content (see p. 6).

Discussion of the Experimental Results

The results of characterising a suspension culture of *Coleus blumei* in CB_2 and CB_4 medium are illustrated in Fig. 5.8. The illustration does not show idealised curves but those obtained when the points are joined up, thus avoiding an arbitrary interpretation of the results.

Growth of *Coleus blumei* cultures starts immediately after inoculation. An obvious lag phase, observed with many other cultures, does not occur. Cell division takes place in both CB_2 and CB_4 media up to the eighth day, when the number of cells starts to fall off as a result of cell death (Fig. 5.8(a)). An increase in dry weight (Fig. 5.8(b)) is observed in CB_2 medium only up to the fifth day and up to the seventh day in CB_4 medium. The wet weight (Fig. 5.8(a)) increases in both media for a slightly longer period because of water storage. All the growth parameters show an abrupt entry into the dying-off phase without an intermediate stationary phase. The growth in medium containing 4% sugar is distinctly better than that in medium containing 2% sugar.

The accumulation of rosmarinic acid (Fig. 5.8(b)) commences only at the end of the growth phase. There is a distinct difference in the two cultures. In medium containing 2% sugar there is scarcely any synthesis of rosmarinic acid, whereas in medium with 4% sugar rosmarinic acid accounts for approximately 9% of the dry weight. In fresh suspension cultures the amount can even reach 20%. The activity of PAL (Fig. 5.8(c)) correlates with the accumulation of rosmarinic acid. An obvious maximum is reached in CB_4 medium on the sixth day of the culture. However PAL activity remains low in CB_2 medium.

The amount of soluble protein in the *Coleus blumei* cells increases significantly during the first three days of culturing, only to fall off continually for the rest of the culture period. The changes in the culture medium occur because the cells take substances from the medium and release others into it. The latter occurs particularly when the cells lyse during the dying-off phase.

The refractive index, a measure of the sugar content of the medium, drops in a linear fashion in both media until sugar is no longer detectable. In CB_2 medium this is the case on the fifth day and in CB_4 medium on the eighth day. The slight increase in the refractive index towards the end of the culture period is due to the cells dying off and releasing their contents into the medium.

The pH value of the medium shows a characteristic drop on the second day of culture only to increase continually thereafter to the end of the culture period (Fig. 5.8(d)). The reason for this is that phosphate is taken up rapidly by the cells (Fig. 5.8(e)) and the medium is deprived of its most important buffering component. Nitrate is taken up by the cells much more slowly (Fig. 5.8(e)). The increase in both phosphate and nitrate at the end of the culture period is again the result of cell lysis.

Conductivity and osmolarity of the medium (Fig. 5.8(f)) mirrors the total amount of ions and osmotically active substances in the medium. They decrease throughout the growth of the culture, and this is to be regarded as a sign of active metabolism. An increase in both these parameters is an indication of an increase in the number of dying cells. The slight increase in osmolarity at the beginning of the culture period in CB_4 medium is caused by sucrose being split into glucose and fructose and thus increasing the number of osmotically active particles.

In general it may be said that suspension cultures of *Coleus blumei* do not show the typical growth pattern of many cell cultures, with lag phase, growth phase,

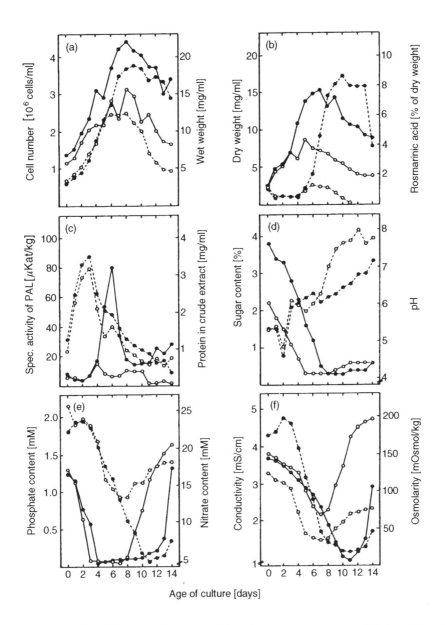

Figure 5.8 Examples of the growth characteristics of a suspension culture of *Coleus blumei* in CB$_2$ (o) and CB$_4$ (●) media over a period of 14 days (taken from Gertlowski, 1991); (a) cell number (---), wet weight of the cells (—); (b) dry weight of the cells (—), rosmarinic acid content of the cells as a function of dry weight (---); (c) specific activity of PAL (—), protein concentration of crude extract (---); (d) sugar content determined from the refractive index (—), pH of the medium (---); (e) phosphate content of the medium (—), nitrate content of the medium (---); (f) conductivity of the medium (—), osmolarity of the medium (---).

stationary phase and dying-off phase. In the *Coleus blumei* cultures there are only two phases to be distinguished: *growth* and *dying-off*. Synthesis and accumulation of rosmarinic acid take place at the end of the growth phase. The amount of rosmarinic acid accumulated is determined by the amount of sugar available at a stage when another nutrient (e.g. phosphate) becomes growth-limiting. Therefore the addition of more sugar or a continual addition of sugar can lead to an increased production of rosmarinic acid.

5.5.3 *Biotransformation of Salicylic Alcohol by a Suspension Culture of* Daucus carota

Principles

Biotransformation is, as well as the biosynthesis of plant products and micropropagation, one of the most interesting areas for potential application of plant *in vitro* culture. *Biotransformation* is defined in biotechnology as the modification of a pure substance with the aid of a living organism or enzymes isolated therefrom. As the result of a single or a few enzyme reactions an improved product is obtained in an increased amount. On the other hand, *biodegradation* is defined as the breakdown of a substrate, whereas the term *bioconversion* is applied to the conversion of a mixture of substrates. It is, however, not always possible to separate the three terms biotransformation, bioconversion and biodegradation (Kieslich, 1984).

The capacity for biotransformation of externally applied substances has already been demonstrated for many plant cell cultures. The initial step for the desired transformation of the substrate is the addition, under sterile conditions, of the organic starting substance to the plant cell culture. During the ensuing incubation the cells convert the substance biochemically into the desired product; depending on the type of cell culture and the substance, the end product is either accumulated in the cells or excreted into the medium.

Substrates may be either substances already available *in vivo* or from another plant species or even chemically synthesised. Biotransformations with plant cell cultures have already been described for many classes of plant secondary metabolites, e.g. for phenols, coumarines, alkaloids or terpenoids. Synthetic xenobiotics such as pentachlorophenol or nitrophenol can be altered by biotransformation in plant cell culture. The reactions leading to the biotransformation, and therefore the reaction products, are determined by the culture used but there is some scope, albeit limited, to influence the results by varying the culture conditions. However, on account of the many cell cultures tested and substrates used, a large number of different biotransformation reactions have been identified, e.g. hydroxylation, reduction of carbonyl groups or carbon double bonds, oxidation of alcohols, esterification of organic acids, glycosidation, methylation, hydrolysis of esters and ethers, epoxylation or isomerisation (for summaries see Reinhard and Alfermann, 1980; Pras, 1990; Suga and Hirata, 1990).

Microbial biotransformations have long been used as production methods, e.g. the production of acetic acid from alcohol by *Acetobacter xylinum*. However the use of plant cell culture for the production of chemical or pharmaceutical substances has not yet established itself. Recently a biotransformation system with commercial potential has been described, the biotransformation of hydroquinone by a suspension culture of

Catharanthus roseus into the β-D-glucoside arbutin (Yokoyama and Yanagi, 1991). Arbutin, used in Asian countries for depigmentation of the skin, can be obtained in high quantities, up to 45% of the dry weight, in suspension cultures of *Catharanthus roseus*. The production costs of the biotransformation with cell culture are roughly the same as those for the chemical synthesis, so there is a chance that the biotransformation may find an industrial application.

The production of glycosides is a typical biosynthetic capacity of the plant cell but one which is much rarer among the microbes. These reactions can be used in biotransformations: for example, to make a nonpolar substance water soluble by the addition of sugar residues or to render substances less volatile through the addition of glycosides and thereby preserve them. The release of the volatile substance can be initiated at a later stage by hydrolysis or a glucosidase reaction.

The present experiment describes such a glycosylation. The substrate is salicylic alcohol (saligenin or 2-hydroxybenzoic acid) and is found in the bark of willow trees. The experimental system is an anthocyanine-containing suspension culture of a variety of carrot, *Daucus carota* L. This suspension culture produces two isomeric β-D-glucosides: salicin (salicylic alcohol-2-O-β-D-glycopyranoside) and isosalicin (salicylic alcohol-1-O-β-D-glycopyranoside). The aim of the experiment is to determine which substances are produced and to what extent the relative amounts of these substances change in the course of cultivation.

Inoculation

Cultures of the cell suspension are grown as 70 ml cultures in 300 ml Erlenmeyer flasks under constant illumination (ca. 75 millilux/m^2 S) on a round shaker. New cultures are set up every week by inoculating 20 ml of a 7-day-old culture with a sawn-off sterile glass pipette into 50 ml 12a medium. Since the biotransformation experiment only requires a small amount of cells the experimental cultures are set up in 100 ml Erlenmeyer flasks in which the volume of fresh medium is 20 ml and the volume of the inoculum is only 8 ml.

Materials Required

Four 300 ml Erlenmeyer flasks each containing 70 ml of a week-old culture of *Daucus carota*; sterile sawn-off and plugged 10 ml glass pipettes; twenty-six 100 ml Erlenmeyer flasks each containing 20 ml 12a medium; one sterile 500 ml Erlenmeyer flask.

Procedure

The four 1-week-old cultures are combined in the 500 ml Erlenmeyer flask and mixed thoroughly. 8 ml aliquots of this suspension are dispensed with the sawn-off pipette into each of the 100 ml Erlenmeyer flasks containing the fresh medium. These flasks are incubated under the conditions described under 'Inoculation'.

Characterisation of Growth Characteristics

In order to compare the biotransformation and the growth characteristics with each other, the growth characteristics must be determined. In general the monitoring of the growth of a suspension culture involves the determination of various parameters. One group of parameters determines the increase in biomass and includes determination of the wet and dry weights or the sink volume. A second group of parameters deals with the use of nutrients. In this case the concentrations of the nutrients in the medium are determined. The different phases of growth (adaptation, growth phase, stationary phase) can be identified with the aid of a small number of parameters. In this experiment it suffices to determine the dry and wet weights of the cells, total sugar concentration, pH and conductivity of the medium.

Procedure

Starting on the day of inoculation, two flasks are used on alternate days to measure the growth parameters of the 'unfed' cultures which act as controls. The methods of determination are given on p. 12–14.

Biotransformation

In order to compare the biotransformation properties of the culture with the growth pattern, the substrate to be transformed is added to cultures of different 'ages'. After a defined period of incubation the suspension is processed. The amount of substrate remaining in the medium is determined. The cells are extracted with an organic solvent, and the product of the biotransformation is determined both qualitatively and quantitatively.

Addition of Substrate

EQUIPMENT

Adjustable automatic pipette capable of delivering 280 μl; sterile pipette tips; 6 ml aqueous, filter-sterilised solution of 100 mM salicylic alcohol.

PROCEDURE

Substrate addition takes place on the 2nd, 6th, 10th and 14th days of culture and must be carried out aseptically. On the above days 280 μl of the salicylic alcohol solution are added to two of the flasks. The final concentration of salicylic alcohol in the culture is 1 mM. To reduce the risk of contamination the shaft of the pipette can be wiped with alcohol immediately before use. The cultures are incubated for 2 days after the addition of salicylic alcohol.

Harvesting and Processing

EQUIPMENT

Freeze-drying apparatus; homogeniser, e.g. Waring blender.

PROCEDURE

At the end of the incubation the cell suspension is treated as for wet and dry weight determinations. The volume of the spent medium is determined with the aid of a measuring cylinder. 500 μl of medium is transferred to a reaction tube which is then stored at $-20\,°C$ until the HPLC analysis is performed. Two 1 g samples of the cells are weighed into test tubes and extracted. The extraction is carried out as follows: 3 ml methanol is added to the cells and the whole is homogenised for 30 s under cooling with ice. The homogenates are centrifuged for 10 min at 5000 g and the supernatant transferred to a graduated test tube with a Pasteur pipette. The precipitate is extracted twice more with 3 ml methanol, the second and third extracts are added to the first, and the total volume is made up to 10 ml.

Analysis

Qualitative Thin-layer Chromatography of the Biotransformation Products

EQUIPMENT

Standard chromatography chamber; 5 μl capillaries; liquid phase: ethyl-acetate : methanol : water = 77 : 13 : 10; tlc plates: silica gel with a fluorescence indicator (e.g. silicagel 60 F_{254}; Merck); sulphuric acid reagent: 5% H_2SO_4 in ethanol; drying oven or thermostatically controlled heating block.

PROCEDURE

500 μl of the methanolic extract is evaporated to dryness and the residue taken up in 50 μl methanol, resulting in a tenfold concentration of the extract. 5–10 μl of this extract is applied to the tlc plate. Salicin, isosalicin and salicylic alcohol are also applied to the plate as reference substances. The chamber is saturated with running buffer and the chromatography is run upwards at room temperature. The chromatograms are viewed under UV light (254 nm) before they are sprayed with sulphuric acid reagent and placed at $110\,°C$ for 10 min. Under these conditions salicylic alcohol compounds are stained reddish-brown and are therefore readily visible. The biotransformation product is identified by its R_f value, its coloration after treatment with sulphuric acid reagent and also by comparison with the reference substances run in parallel. The R_f values in this chromatography system are approximately 0.70 (salicylic alcohol), 0.3 (salicin) and 0.35 (isosalicin).

Quantitative HPLC Analysis

EQUIPMENT

HPLC with gradient facility, UV detector and integrator; Column: Nucleosil 100 5C18, 250×4.8 mm (Macherey and Nagel) or an equivalent RP18 column; elution buffer and solvent: 0.5% H_3PO_4 in H_2O (running buffer A) and acetonitrile (running buffer B); gradient: $t = 0$ min: 15% B, flow rate: $t = 0$ min: 0.7 ml/min; $t = 10$ min: 35% B, $t = 10$ min: 1.0 ml/min; $t = 14$ min: 47% B, $t = 15$ min: 1.3 ml/min; $t = 15$ min: 15% B, $t = 18$ min: 15% B, UV detector set at 225 nm.

PROCEDURE

All the samples are thawed and carefully mixed before centrifuging for 10 min at 10 000 g to remove any cell debris. The supernatant can be used directly for HPLC analysis.

For the quantitative analysis, 100 μl of the extract is evaporated to dryness and then taken up in 100 μl H$_2$O. Any insoluble substances are removed by centrifugation for 10 min at 10 000 g. The aqueous supernatant is used directly for HPLC analysis.

After separation of the individual substances on the HPLC column they can be identified with the UV detector. The retention time is diagnostic for each substance. The concentration of each substance in the extracts can be calculated from the height or area of the relevant peak, ideally using the integrator. The basis for the calculation is a defined amount of the pure substance which has been analysed and quantified by HPLC.

Discussion of the Results

The products of the biotransformation can be identified by means of thin-layer chromatography. HPLC analysis then allows the amounts of substrate and product(s) to be quantified so that the uptake and metabolism of the substrate as well as the product formation can be determined. Furthermore the amount of substrate remaining and the product yield can be calculated. The rate of conversion can be determined if one takes into account incubation time and the number of cells.

The determination of wet and dry weights of the control cultures permits an approximate characterisation of the growth of the culture so that information can be obtained about the stage of the growth curve for the addition of salicylic alcohol to obtain good rates of substrate turnover and product formation.

Fig. 5.9 illustrates the HPLC tracing of a standard solution obtained with the chromatographic conditions described above. Figs 5.10 and 5.11 illustrate the growth curves of the cells obtained by determining wet and dry weights and the amounts of the derivatives of salicylic alcohol obtained after various 2-day incubation periods.

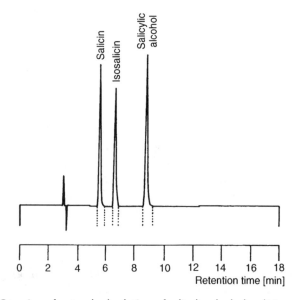

Figure 5.9   HPLC tracing of a standard solution of salicylic alcohol, salicin and isosalicin.

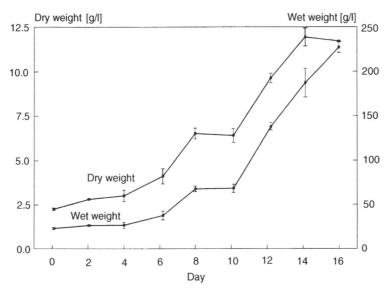

Figure 5.10 Growth curve of a suspension culture of *Daucus carota* drawn up using the dry and wet weights of the cells.

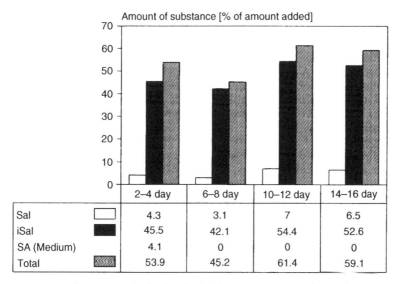

		2–4 day	6–8 day	10–12 day	14–16 day
Sal	☐	4.3	3.1	7	6.5
iSal	■	45.5	42.1	54.4	52.6
SA (Medium)		4.1	0	0	0
Total	▨	53.9	45.2	61.4	59.1

Figure 5.11 Biotransformation of salicylic alcohol by a suspension culture of *Daucus carota*: correlation between the age of the culture and its capacity for biotransformation.

Variations

This experiment describes an investigation into the glucosidation of salicylic alcohol to salicin and isosalicin as a function of the age of the culture. In the same way, other substrates or glucosidation reactions can be looked at: e.g. glucosidation of salicylic acid to 2-O-salicylic acid-β-D-glucoside or the transformation of salicylaldehyde into the glucosides of salicylic alcohol and salicylic acid. In the last instance, helicin

229

(salicyl aldehyde-2-O-β-D-glucopyranoside), the direct glucosidation product of salicyl aldehyde, cannot be determined. When trying out these other possibilities it should be remembered that the use of different substrates may necessitate an alteration in the gradient programme so that the relevant aglyca and glycosides are eluted.

Reduction of the timespan of the experiment demands, for example, an investigation into the dependency of the biotransformation on the substrate concentration or the length of incubation. For such experiments the substrate could be added in different concentrations at a specific stage of the growth curve – suitable time points are either late in the growth phase or at the beginning of the stationary phase – so that the correlation between the amount of glucoside obtained and the substrate concentration may be determined. Another possibility is to remove samples at regular times after the addition of the substrate and so determine the length of incubation required to give the maximum amount of product.

References

BAJAJ, Y.P.S. (1985 ff.) *Biotechnology in agriculture and forestry* (Berlin: Springer Verlag).

BHOJWANI, S.S. and RAZDAN, M.K. (1983) *Plant tissue culture: theory and practice* (Amsterdam: Elsevier).

BONGA, J.M. and DURZAN, D.J. (1987) *Cell and tissue culture in forestry*, Vols 1–3 (Dordrecht: Martinus Nijhoff).

DIXON, R.A. (ed.) (1985) *Plant cell culture: a practical approach* (Oxford: IRL Press).

DODDS, J.H. and ROBERTS, L.W. (1985) *Experiments in plant tissue culture*, 2nd edn (Cambridge: Cambridge University Press).

EVANS, D. *et al.* (eds) (1983 ff.) *Handbook of plant cell culture* (New York: Macmillan).

FOWLER, M.W. and WARREN, G.S. (eds) (1992) *Plant biotechnology* (*Comprehensive biotechnology*, Second Supplement) (Oxford: Pergamon).

GAMBORG, O.L., MILLER, R.A. and OJIMA, K. (1968) Nutrient requirements of suspension cultures of soybean root cells. *Exp. Cell Res.* **50**, 151–8.

GEORGE, E.F., PUTTOCK, D.J.M. and GEORGE, H.J. (1987) *Plant culture media*, Vol. 1: *Formulation and uses* (Edington: Exegetics Ltd).

GEORGE, E.F., PUTTOCK, D.J.M. and GEORGE, H.J. (1988) *Plant culture media*, Vol. 2: *Commentary and analysis* (Edington: Exegetics Ltd).

GEORGE, E.F. and SHERRINGTON, P.D. (1984) *Plant propagation by tissue culture. Handbook and directory of commercial laboratories* (Edington: Exegetics Ltd).

GERTLOWSKI, C. (1991) Einfluß des Kulturmediums auf die Production von Rosmarinsäure in Suspensionkulturen von *Coleus blumei*. Diploma thesis, University of Düsseldorf.

HESS, D. (1992) *Biotechnologie der Pflanzen* (Stuttgart: Eugen Ulmer).

KIESLICH, K. (1984) Introduction. In: H.-J. Rehm and G. Reed (eds) *Biotechnology – a comprehensive treatise in 8 volumes,* Vol. 6a: *Biotransformations*, pp. 1–4 (Weinheim: VCH).

MURASHIGE, T. and SKOOG, F. (1962) A revised medium for rapid growth and bioassays with tobacco tissue cultures. *Physiol. Plantarum* **15**, 473–97.

PETERSEN, M. (1993) *Colues spp.: in vitro* culture and the production of forskolin and rosmarinic acid. In: Y.P.S. Bajaj (ed.) *Biotechnology in agriculture and forestry*, Vol. 26, pp. 69–92 (Berlin: Springer Verlag).

PRAS, N. (1990) Bioconversion of precursors occurring in plants and of related synthetic compounds. In: H.J.J. Nijkamp, L.H.W. van der Plas and J. van Aartijk (eds) *Progress in plant cellular and molecular biology*, pp. 640–9 (Amsterdam: Kluwer Academic Publications).

REINHARD, E. and ALFERMANN, A.W. (1980) Biotransformation by plant cell cultures. *Adv. Biochem. Eng.* **16**, 49–84.

SEITZ, H.U., SEITZ, H. and ALFERMANN, A.W. (1985) *Pflanzliche Gewebekultur – ein Praktikum* (Stuttgart: Gustav Fischer Verlag).

SUGA, T. and HIRATA, T. (1990) Biotransformation of exogenous substrates by plant cell cultures. *Phytochemistry* **29**, 2393–406.

VASIL, I. (ed.) (1984 ff.) *Cell culture and somatic cell genetics of plants* (Orlando: Academic Press).

YOKOYAMA, M. and YANAGI, M. (1991) High-level production of arbutin by biotransformation. In: A. Komamine, M. Misawa and M. di Cosmo (eds) *Plant cell culture in Japan*, pp. 79–91 (Tokyo: CMC Co.).

6

Statistical Planning and Analysis of Experiments, and Scaling-up

K.-H. WOLF

6.1 SYMBOLS USED

Symbol Definition		Unit
A_L	area of a section of an acentric cylinder	m^2
A_R	area of free movement through the section, $A_R = \pi(d_R^2 - d_w^2)/4$	m^2
a_1	distance of the neutral thread of the flow cylinder from the container wall	m
B	determinant	–
B^*	corrected determinant	–
b_i	coefficient in Eqn (6.3)	as given in the text
C	coefficient in Eqn (6.22)	–
c_i	concentration	$kg\ m^{-3}$
D_{ax}	coefficient of effective axial diffusion	$m^2\ s^{-1}$
d_1	internal diameter of the vessel	m
d_2	diameter of the stirrer	m
d_3	diameter of the flow cylinder	m
d_R	internal diameter of the section	m
d_{sp}	equivalent slit diameter $d_{sp} = [4/\pi(A_L n_L + A_R)]^{1/2}$	m
d_w	diameter of the stirred cylinder	m
E	activation energy	$J\ mol^{-1}$
e	acentric stirred area	m
h_0	working height (non-aerated)	m
h_1	height of stirrer blades	m
h_2	height of stirring unit	m
K_I	inhibition constant	$kg\ m^{-3}$

(continued)

233

Continued

Symbol	Definition	Unit
K_s	Monod constant; substrate conc^n at which half μ_{max} is reached	kg m^{-3}
m	number of factors in the planned expt.	–
m_s	maintenance coefficient	s^{-1}
N	number of experimental points	
N_s	number of the vertical current breaker with the diameter d_3	–
n	speed of the stirrer	s^{-1}
n_i	exponential in Eqn (6.20)	–
n_L	number of acentric holes in the section	–
P	stirring power	kg m^2 s^{-3}
p	number of constants or polynomial degree	–
$q_i = \dfrac{R_i}{c_x}$	specific mass change velocity	s^{-1}
R_i	mass change velocity	kg m^{-3} s^{-1}
T	temperature	K
t_R	duration of reaction	s
v	coefficient of reaction	–
w	average flow speed	m s^{-1}
$Y_{j/i}$	coefficient of yield with respect to product j and substrate i	kg kg^{-1}

Indices

i	reaction component, current number
j	reaction product, current number
L	fluid
max	maximum
P, p	product
S, s	substrate
X, x	biomass

Dimensionless Constants

$$Be = c_w N_s \frac{d_3}{d_1} \frac{h_3}{d_1} \qquad \text{Bewehrungs number}$$

$$Bo = \frac{w h_0}{D_{ax}} \qquad \text{Bodenstein number}$$

$$Ne = \frac{P}{\rho n^3 d_2^5} \qquad \text{Newton number, power number}$$

$$N_{st} = \frac{nd_2}{w}$$ Strouhal number

$$Re = \frac{wd_1}{v}$$ Reynolds number for laminar flow

$$Re = \frac{nd_2^2}{v}$$ Reynolds number for turbulent flow

$$\pi_i$$ General dimensionless number

Greek Characters

Θ	temperature	^0C
η	dynamic viscosity	$kg\ m^{-1}s^{-1}$
μ	specific growth rate	s^{-1}
v	kinematic viscosity	$m^2\ s^{-1}$
ρ	density	$kg\ m^{-3}$

6.2 STATISTICAL PLANNING OF EXPERIMENTS

6.2.1 Basic Principles

In the last 30 years scientific theory has caused a complete change in the planning and execution of experiments. Previously an experiment was regarded as a deterministic procedure. The classic concept is that $n - 1$ factors (where $n \geq 2$) were kept constant and that one factor after another was altered to determine the relationship between the result and the variable. Conversely the theory of modern experimentation is based on the following principles.

1 The experiment is examined from the viewpoint of mathematical statistics.

2 Mathematical statistics provide the means of evaluating and analysing the results.

3 The maximum use is made of the amount of variation in the factors (variables), i.e. the necessary precision in the solution is obtained from the minimum number of experiments.

4 The experimental theory offers a logical strategy, relevant methods of solving the problems and sound information about the reliability of the experiments.

The modern concept is known as *statistical experimental planning*. All the methods of statistical experimental planning allow the simultaneous variation of several factors or variables according to a specific plan.

Statistical experimental planning has found various applications in bioprocessing. The most important are:

- optimisation of enzyme activity in relation to the composition of the nutrient medium

- optimisation of the enzyme yield in relation to the production parameters of the fermentation, e.g. T, $\dot{V}_G$, n.

Statistical planning of experiments involves working with standardised plans or experimental programmes which can be expanded as necessary, e.g. transfer from a linear to a quadratic model. Information already obtained will be taken into consideration.

Transformation of the changes in the variables to $+1$, -1 and 0 permits the drawing up of statistical experimental plans independently of the problem to be solved and a simplified calculation of the estimated coefficients. This is particularly pertinent to the so-called *orthogonal plans*, i.e. in which the product matrix on the left side of the normal equation system is a diagonal matrix.

After the factors (x_i) and the expected values (y_i) have been specified, the model can be set up and the experiment started. The following steps have to be dealt with (see Fig. 6.1):

1. postulation of a model plan
2. planning of the experiment
3. performance of the experiment
4. evaluation of the results
5. statistical analysis of the calculated model.

The steps 1, 2 and 4 will be described only briefly here. The model proposition $f(x)$ can, in principle, be chosen arbitrarily. If the assumption is made that the real (unknown) correlation between the initial value y and the factors x is:

$$y = \phi(x) \tag{6.1}$$

then $\phi(x)$ must be an analytical, multiply-differentiated function. At the point $x = 0$ (this is always attainable by transformation) Eqn (6.1) develops as a Taylor series:

$$y = y_{x=0} + \sum_{i=1}^{n} \frac{\partial \phi}{\partial x_i}\bigg|_{x=0} x_i + \frac{1}{2} \sum_{\substack{i=1 \\ j=1 \\ i=j}}^{n} \frac{\partial^2 \phi}{\partial x_i \partial x_j}\bigg|_{x=0} x_i x_j + \frac{1}{2} \sum_{i=1}^{n} \frac{\partial^2 \phi}{\partial x_i^2}\bigg|_{x=0} x_i^2 + \cdots \tag{6.2}$$

which will be ended after the terms of the second order. This gives rise to the polynomial expression:

$$y = b_0 + \sum_{i=1}^{n} b_i x_i + \sum_{\substack{i=1 \\ j=1 \\ i=j}}^{n} b_{ij} x_i x_j + \sum_{i=j}^{n} b_{ii} x_i^2 \tag{6.3}$$

or

$$y = b_0 + \sum_{i=1}^{n} b_i f_i(x_i, \ldots, x_n) + e \tag{6.4}$$

This polynomial model is an approximation of the real (but unknown) correlation, where the degree of coincidence can be influenced effectively by the degree of

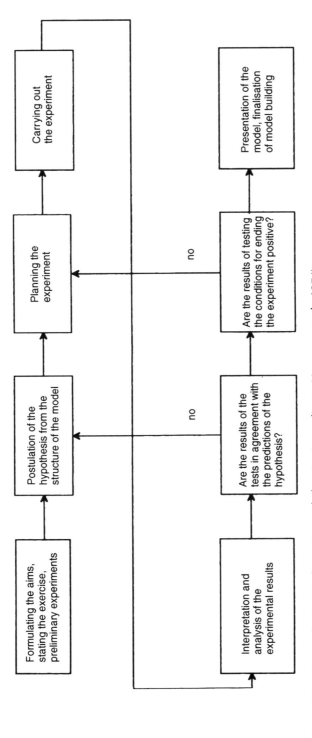

Figure 6.1   Strategy for statistical experimental planning (according to Hartmann *et al.*, 1974).

the polynomial (i.e. its structure). Since statistical models give only a restricted description of $\phi(x)$ (i.e they are only valid in the immediate proximity of the development point x), one is restricted mostly to linear and quadratic expressions, in rare instances to cubic polynomials. This depends in particular on the large number of coefficients $(b+1)$ in the polynomial model and increases with the degree of the polynomial r:

$$b = \binom{i+r}{r} \tag{6.5}$$

As an example, when $i = 3$ and $r = 2$ there are 10 coefficients b_i (i = the number of variables).

In the following, only polynomials of the type given in Eqn (6.3) are used, although in principle other equations can be used, e.g. exponential forms.

The principle behind the calculation of the experimental values according to Eqn (6.3) or generally according to Eqn (6.4) is based on the following considerations. All experimental values

$$x_i = (x_{i1}, x_{i2}, \ldots, x_{in}) \quad i = 1, \ldots, N \tag{6.6}$$

are summarised as plan matrix X termed the experiment plan

$$X = \begin{bmatrix} x_{11} & x_{12} \cdots x_{1n} \\ \vdots & \vdots \quad \vdots \\ x_{N1} & x_{N2} \quad x_{Nn} \end{bmatrix} \tag{6.7}$$

The point x_0 is the centre of the experimental plan. A central experimental plan exists if its mid-point lies at the origin of the coordinates at $x = 0$. By means of linear transformation any experimental plan can be centred.

The range of the possible values of the factors is known as the range of the experimental plan. All points x_i of the range must lie within it. The range of the experimental plan can be determined by inequalities:

$$-1 \leqslant x_j \leqslant 1 \quad j = 1, \ldots, n \tag{6.8}$$

(For partial factor plans, e.g. Box–Hunter (Box and Hunter, 1965) the transformed values of x_i of the inequality can be >1).

On account of the conditions imposed by Eqn (6.8) the range of the experimental plan is represented by a hypercube. The boundary values of the variables are -1 and $+1$, i.e. the variables are normalised. The standardisation of the variables on the natural scale x_i^* is done according to the following formula:

$$x_i = x_i^* - \frac{x_{i,\max}^* + x_{i,\min}^*}{2} \bigg/ \frac{x_{i,\max}^* - x_{i,\min}^*}{2} \tag{6.9}$$

where $x_{i,\min}^*$ and $x_{i,\max}^*$ are the postulated minimum and maximum values of the relevant variable i of the range in the experimental plan. The magnitude of the difference $[\Delta x_i]$ of the natural scale x_i^* from the minimum value $x_{i,\min}^*$ or the maximum value $x_{i,\max}^*$ has to be equal on the grounds of symmetry.

6.2.2 *Planned Experiments – Setting-up and Calculation*

The sum of all values of x whose coordinates are $+1$ $(+)$ or -1 $(-)$ is termed a complete 2^m variable plan. It consists of

$$N = 2^m \tag{6.10}$$

experimental values.

If $m = 2$, 2^m is always $>(m + 1)$ and the method of least squares gives the number of experiments required for estimation of the parameters.

For a wide range of variables a linear model is often insufficient. Therefore one must take into account at least the quadratic terms of the Taylor progression (Eqn (6.3)).

The correlation existing between the number of parameters in different models and the number of measurements at various stages of the experimental plan is shown in Table 6.1. For a 2^m variable plan every variable is assigned two values and all combinations are taken into account.

The calculation is particularly simple if one does not use physical numbers for x_1, x_2, ... x_m but substitutes, after transformation according to Eqn (6.9), values of -1 or $+1$ for x_1.

For example when $m = 2$, according to the rules of calculating the regression, one arrives at the normal equation system:

$$N b_0 + b_1 \sum x_{1i} + b_2 \sum x_{2i} = \sum y_i$$
$$b_0 \sum x_{1i} + b_1 \sum x_{1i}^2 + b_2 \sum x_{1i} x_{2i} = \sum x_{1i} y_i \tag{6.11}$$
$$b_0 \sum y_{2i} + b_1 \sum x_{1i} x_{2i} + b_2 \sum x_{2i}^2 = \sum x_{2i} y_i$$

Table 6.1 Parameters for different models and number of observed values obtained with various experimental plans.

Parameter				Number of observed values		
No. of factors	Linear	Quadratic	Cubic	Box–Wilson Plan (Eqn (6.10))	3^m Plan	Box–Hunter Plan (Eqn (6.11))
m	$m+1$	$\binom{m+2}{2}$	$\binom{m+3}{3}$	2^m	3^m	$2^m + 2m + n_c$
2	3	6	10	4	9	13
3	4	10	20	8	27	20
4	5	15	35	16	81	31
5	6	21	56	32	243	52
6	7	28	84	64	729	91
7	8	36	120	128	2187	163
8	9	45	165	256	6561	
9	10	55	220	512	19 684	
10	11	66	286	1024	59 049	

where N is the number of measurements taken. Using the normalised values, Eqn (6.11) simplifies itself to:

$$4b_0 = y_1 + y_2 + y_3 + y_4$$
$$4b_1 = -y_1 - y_2 + y_3 + y_4 \qquad (6.12)$$
$$4b_2 = -y_1 + y_2 - y_3 + y_4$$

The determination of the coefficient b_i from Eqn (6.12) is elementary, and one immediately arrives at the regression equation:

$$y = b_0 + b_1 x_1 + b_2 x_2 \qquad (6.13)$$

One can now make a series of statistical calculations to evaluate the fit to:

- coefficient of correlation r
- determinant $B \ [r^2 = B]$
- corrected determinant B^*
- remaining squared sum RSS
- remaining variation S_R
- coefficient of variation v

At least the determinant has to be calculated.

As m increases, the number of degrees of freedom increases so rapidly that not all 2^m experiments have to be carried out. One can set up *partial plans* (often designated as 2^{m-1}, 2^{m-2}, ... plans). Tables for multidimensional plans are to be found in the literature. If one designs them oneself it should be remembered that the plan has to be orthogonal.

The experimental effort increases considerably with increasing numbers of variables (Table 6.2). The numerical effort necessary for determining the coefficients, the statistical constants and tests increases so much that it can only be dealt with using a computer and the appropriate software. As an example of a software package for the calculation of the results of statistically planned experiments, STATMOD (developed at and available from Institut für Mechanische Verfahrenstechnik und Systemverfahrenstechnik, Technical University, Dresden) may be recommended. The software for empirical statistical modelling offers a variety of methods for data analysis and modelling:

- calculation of empirical statistical values
- component analysis
- factor analysis

Table 6.2 Transformed experimental values for a 2^m factor plan when $m = 2$.

i	x_{1i}	x_{2i}	x_{1i}^2	$x_{1i}x_{2i}$	x_{2i}^2	y_i	$x_{1i}y_i$	$x_{2i}y_i$
1	-1	-1	$+1$	$+1$	$+1$	y_1	$-y_1$	$-y_1$
2	-1	$+1$	$+1$	-1	$+1$	y_2	$-y_2$	$+y_2$
3	$+1$	-1	$+1$	-1	$+1$	y_3	$+y_3$	$-y_3$
4	$+1$	$+1$	$+1$	$+1$	$+1$	y_4	$+y_4$	$+y_4$
Σ	0	0	4	0	4			

- quasi-linear multiple regression
- class regression
- global and collective cluster analysis methods
- discrimination analysis, adaptation classification
- fuzzy classification, unfocused production regulation systems
- components of protocols
- various graphic presentations.

The programme is dialogue-oriented and menu-steered; the help function supports the user in the choice of modelling methods suitable for the data and the problem being investigated, the choice of parameters and the interpretation of the data. *Hardware required*: IBM-PC/XT/AT compatible with MS-DOS (at least version 3.0), at least 640 kB memory capacity, graphic program and mouse.

6.2.3 *Limits of Statistical Experimental Planning*

In spite of its many advantages, statistical experimental planning also has its limitations. The results and/or functional correlations obtained are not scientifically proven facts but represent only formal equations valid under certain conditions.

- Conclusions drawn cannot be applied to a scaled-up version of the experiment.
- Extrapolation outside the limits set for the experiment are invalid.
- The coefficients of correlation and/or determinants obtained are often satisfactory only with non-linear regression equations or in complicated planned experiments.

6.3 INTERPRETATION OF EXPERIMENTS – COMPUTER-ASSISTED TECHNIQUES

An effective interpretation of the data obtained in biochemical and microbiological processes is possible with a PC and the appropriate software. In the field of bioprocessing, the following software systems are particularly useful:

- Modellbank Biotechnologie
- BIOMOD.

Both these systems were developed in university institutes and have been taken on board by the biotechnology industry. Both systems have essentially the same capacity, however Modellbank Biotechnologie is recommended for the following reasons:

- advantages in the use of the menu
- constant updating of the software by the developer
- its integration into teaching at about 25 German universities and some 400 users involved in research.

Both Braun Diessel Biotech GmbH and Münzer + Diehl Electronic GmbH use the integrated Modellbank Biotechnologie as part of their fermenter control system (micro-MFCS-modelling system). The hardware requirements of the system are: IBM/XT or compatible equipment; EGA or VGA graphic cards; at least 512 kbyte RAM; numeric co-processor. Any 386 PC can be used for running the program.

If the Modellbank Biotechnologie is not integrated into a fermenter control system it is capable of *interpretation* (correcting parameters to fit a given model), *simulation* and *optimisation* by means of simulation. Once installed, the Modellbank is easily used even by those with no knowledge of programming, because of the menu and the many help texts. The software system is made up of the following modules:

- data editor
- enzyme kinetics
- growth and production
- user function
- differential equations (up to a maximum of 10)
- multidimensional regression
- approximation
- interpretation
- installation aid
- options
- help menus.

The data editor operates in the full-screen mode and is capable of accepting and delivering data from other programs or data collection systems in different formats (dBase, Text). The data editor provides, in addition to the usual functions, a number of possibilities, e.g.

- transformation according to a given formula
- evenning-out of data
- graphical presentation.

In particular, *transformation* is important for the interpretation of the following graphical representation. First of all, deductions R_i (data editor A) of the state variables will be calculated from the data sets available using the difference approximation. Interdependent constants with respect to time can then be calculated from the data sets available:

μ specific growth rate, specific rate of division

q_s specific rate of breakdown

$Y_{X/S}$ yield coefficient.

Using the data editor (Z-drawing) one can obtain appropriate representations of:

$\mu \quad = f(C_S)$

$\mu \quad = f(C_X)$

$q_s \quad = f(\mu)$

$Y_{X/S} = f(\mu)$ or $1/Y_{X/S} = f(1/\mu)$

$R_S \quad = f(R_X)$

all of which are important for model syntheses, e.g.

1 rate of synthesis

2 dependency on stoichiometry.

For specific problems which cannot be dealt with by the defined models, the Modellbank has a so-called 'user function'. This module is available for user-specific functions. The various models may be stored in different files so that once they are defined they are always available. If the experimental values are not prepared according to specific instructions it is relatively difficult to judge the initial values. With initial values which are very different from the optimum and not in the correct combination the analysis results in:

- very long periods of calculation, in particular with digital systems, or
- failure of the system, or
- a so-called local optimum.

The values given in Table 6.3 may be taken as estimates of the approximate values for the *maximum specific growth rate* or the *maximum specific rate of division*. Examples of the specific growth rate μ_{max} for selected microorganisms are given in Table 6.4.

For determination of the temperature dependence a reasonable estimate of the activation energy is required since this quantity appears in the exponential term.

Table 6.3 Range of the kinetic constants μ_{max} or ν_{max} of different microorganisms according to Diekmann and Metz (1991).

Microorganism	μ_{max} (h^{-1}) or ν_{max}
hyphal fungi	0.1–0.34
yeasts	0.34–0.6
bacteria	0.69–3.0

Table 6.4 Kinetic constants of selected microorganisms.

Organism	Θ (°C)	μ_{max} (h^{-1})
Aspergillus niger	30	0.2
Aspergillus nidulans	20	0.09
	25	0.148
	30	0.215
Penicillium chrysogenum	25	0.123
Streptococcus equisimilis	30	0.74
Saccharomyces carlsbergensis	20	0.12
Claviceps purpurea	24	0.046

Table 6.5 gives estimates of the activation energies of several different reactions. More accurate estimates of the activation energy for microbial growth are given in Table 6.6. Further information for specific cases is to be found in Atkinson and Mavituna (1983) and Rehm and Reed (1985).

If the equations for the concept of energy maintenance are to be used in modelling, one requires an estimate of the maintenance coefficient m_s. In general,

$$10^{-4} < m_s/\text{h}^{-1} < 10^{-1} \tag{6.14}$$

Table 6.5 Activation energy, E, of various processes.

E (10^3 J/mol)	Process
1–50	Physical properties
8–30	Diffusion in foodstuffs
10–100	Enzyme reactions; microbiological reactions
50–150	Chemical reactions
100	Thiamine breakdown
80–150	Maillard reaction
70–200	Enzyme inactivation
150–300	Irreversible protein inactivation
170–250	Killing of microorganisms (vegetative forms)
270–500	Killing of microorganisms

Table 6.6 Activation energy of selected microorganisms.

Organism	Temperature range (°C)	E (J/mol)	Reference
Candida kefyr	35–40 (yeast)	4.4×10^4	Hamad (1986)
Hansenula polymorpha	35–40 (yeast)	1.0×10^5	Hamad (1986)
Saccharomyces carlsbergensis	10–25 (yeast)	7.1×10^4	Wolf and Voigt (1992)
Pophyridium crentum	20–30 (microalgae)	8.0×10^4	Wolf (1991)
Pediococcus acidolacticus	20–45 (bacterium)	5.2×10^4	Wolf and Voigt (1992)
Enterococcus faecium	20–40 (bacterium)	3.19×10^4	Wolf and Voigt (1992)
E. coli psychrophilic	12–26 (bacterium)	1.2×10^5	Pirt (1975)
Pseudomonas spp.	12–30 (bacterium)	5.3×10^4	Pirt (1975)
Klebsiella aerogenes	20–40 (yeast)	6.0×10^4	Pirt (1975)
Cells from mouse tissue	31–38 (animal cells)	1.2×10^5	Pirt (1975)
Aspergillus nidulans	20–37 (fungus)	5.9×10^4	Pirt (1975)
Serpula lacrymans	20–38 (fungus)	9.15×10^4	Körner *et al.* (1991)
F. pinicola	16–24 (fungus)	4.89×10^4	Körner *et al.* (1991)
C. puteana	20–38 (fungus)	5.46×10^4	Körner *et al.* (1991)

To estimate the substrate constant K_S one may assume

$$K_S \approx \frac{C_{S,max}}{10} \qquad (6.15)$$

and in the case of substrate inhibition

$$K_I \approx 5K_S = \frac{C_{S,max}}{2} \qquad (6.16)$$

For an estimation of the initial value of the yield coefficient the following information can be used:

$$0.2 \leqslant Y_{X/S} \lesssim 1.7; \ Y_{X/S} \text{ in } g_X/g_S \qquad (6.17)$$

$$0.5 \leqslant Y_{X/0} \lesssim 3.5; \ Y_{X/0} \text{ in } g_X/g_0 \qquad (6.18)$$

$$0.01 \leqslant Y_{X/Q} \lesssim 0.1; \ Y_{X/Q} \text{ in } g_X/kJ \qquad (6.19)$$

6.4 SCALE-UP

6.4.1 *Basic Principles*

The complex basic processes of bioprocessing are determined by several variables, classified as follows:

1   working parameters
2   geometrical constants
3   material properties.

For simple processes the determination of the model structure is just possible; however, the calculation of the model parameters or the mathematical solution of the equations of the model system is particularly complicated if not impossible. Multiphase processes in real reactors where substances are altered or converted are usually expressed only in terms of their basic structure or in general terms. The structures are unknown. Similarity theory has been found useful for solving several kinds of these problems, in particular fluid motion, heat and material transport. This theory is based on scientifically sound model structures and leads, after restatement with the introduction of dimensionless constants, to critical equations whose free parameters can be determined numerically after the evaluation of specific experiments. One special method, which combines model formulation and solution, is dimension analysis. From the equations obtained, this type of analysis is related to similarity theory and some of the steps involved are the same (Fig. 6.2) although the premises are quite different.

The dimensional analysis is used if the information available is not as complete as it could be. Dimensional analysis requires no information about the model structure since the model is constructed by calculating the variables and their units as well as realisation of constants and critical equations in analogy to the similarity theory. The critical equations have the following form:

$$\pi_n = K \ \pi^{n_1} \pi^{n_2} \dots \pi^{n_x} \qquad (6.20)$$

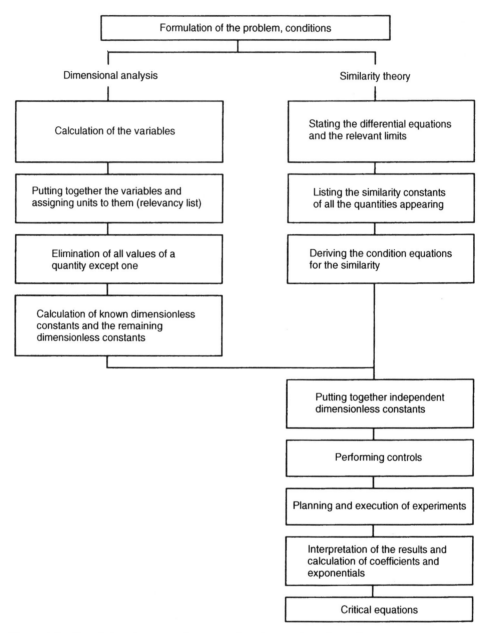

Figure 6.2   Correlations between dimensional analysis and similarity theory.

where π^n is the dimensionless dependent variable and π^{n_i} ($i = 1, \ldots, x$) are the independent dimensionless variables. Whether the critical equation is obtained by similarity theory or dimensionless analysis has no effect on the model.

The way from variables with dimensions to dimensionless constants is by means of dimensional analysis attributed to Lord Rayleigh (1915) and based on the fundamentals of Buckingham's π theorem (1914). There are two methods for obtaining constants:

1   Rayleigh method

2   Buckingham method.

For instance the *Navier–Stokes equations* can be stated without dimensions:

$$\frac{\partial W_i}{\partial (Ho)} + \sum_k W_k \frac{\partial W_i}{\partial X_k} = -\frac{\partial (Eu)}{\partial X_i} + \frac{1}{Re} \sum \frac{\partial^2 W_i}{\partial X_k^2} + \frac{1}{Fr} \tag{6.21}$$

where

$$X = \frac{x_i}{L}, \qquad W = \frac{w_i}{U_0}$$

$$Ho = \frac{U_0 \tau}{L} \qquad \text{(simultaneous number)}$$

$$Eu = \frac{P}{U_0^2} \qquad \text{(Eulerian number)}$$

$$Re = \frac{U_0 L}{\eta} \qquad \text{(Reynold's number)}$$

$$Fr = \frac{U_0^2}{gL} \qquad \text{(Froude's number)}$$

where g is gravitational acceleration. The dimensionless coefficients (criteria of similarity) of the Navier–Stokes equation are:

Ho, Eu, Re, Fr.

The functional correlation of the constants, which reflects the physical and temporal, i.e. kinematic and dynamic, similarity to flow properties:

$\mathrm{f}(Eu, Ho, Re, Fr) = 0$

is often determined experimentally for certain values, and in this case the geometric constant (Γ_i) for geometric similarity has to be included:

$\mathrm{f}(Eu, Ho, Re, Fr, \Gamma_i) = 0.$

This correlation represents the similarity theory's solution of the equation system. The significance of deriving the critical equation from a differential equation has the advantage that no variable is left out.

   In line with Eqn (6.20) the critical equation is as follows when it is expressed in terms of the constant:

$$Eu = CHo^{n_1} Re^{n_2} Fr^{n_3} \Gamma^{n_4} \tag{6.22}$$

The advantages of similarity theory which implicates dimension analysis are as follows:

1   reduction in the number of parameters describing the problem (π theorem);

number of dimensionless constants = number of variables minus the row of the unit matrix $(s = n - r)$

2 accurate scale-up

3 emphasis of the physical process

4 more freedom in the choice of the variables and accurate extrapolation of single parameters (within the limits)

5 independence from a dimensional system.

Regarding point No. 2, it must be said that an absolutely accurate scale-up is only possible if all conditions of similarity are fulfilled. The criteria are as follows:

- geometrical similarity
- hydrodynamic similarity (mechanical similarity)
- thermal similarity
- chemical similarity.

It can be shown that a model of physico-chemical correlation only allows a complete description of partial processes. There is no consideration of this for the three sets of results. One must therefore accept a partial similarity which is determined by the operation ruling the process. For reaction processes chemical similarity has top priority.

Scaling-up where only hydrodynamic or energy similarity is required is virtually always possible with the aid of similarity theory.

Principles and applications of the method of dimensional analysis are to be found in the literature.

6.4.2 Calculation of a Critical Equation – π Relation

The following approach is recommended by Zlokarnik (1991).

1 List of all relevant variables (relevancy list):
    (i) a single constant term (dependent variable)
    (ii) all independent variables (process parameters, geometric constants and material constants).

2 Determination of a complete set of constants:
    (i) setting up the dimension matrix
    (ii) calculation of the range r of the matrix (Gauss algorithm)
    (iii) formulation and, if necessary, re-definition and linking of the constants.

3 Reduction in the number of constants after consideration of other relevant relationships. (A statistical experimental plan could also be used here under certain circumstances.)

Further suggestions regarding the preparation of a relevancy list, again from Zlokarnik (1991), are given below.

1 A relevancy list must be made for every constant. (Each problem investigated requires a new or extra relevancy list.)

2 The relevancy list often becomes overloaded and complicated when all geometric

constants are included. The introduction of a single length measurement allows the remaining geometric values to be rendered dimensionless (Γ_i).

3   Only those variables which are independent of each other should be listed (not ν but ρ and η).

4   The variables listed must be measurable under the conditions.

5   Natural constants (gas constant, g) must appear in the list if they affect the process.

If the variables are known from diagonals or *a priori*, the following procedure should be taken.

1   Make a relevancy list.

2   Construct a dimension matrix (divided into nuclear and remainder matrix).

3   Carry out linear transformations of the lines to produce a unit matrix and a remainder matrix.

4   Read the values of π or the corresponding dimensionless constants.

5   Reformulate or recombine the π values into known constants.

Zlokarnik's textbook *Dimensional analysis and scale-up in chemical engineering* is an excellent source of information and examples.

The method is described below for a simple example.

Example

The example deals with axial diffusion (back-mixing) of a gas-free homogenous fluid in a stirred and divided Tower-type fermenter. Axial diffusion rate for mixing $\beta = \dot{V}_R / \dot{V}$ between two stages. N_s is the number of the flowbreaker, which can only be included in connection with the diameter of the flowbreaker. Since N_s is not variable but is nevertheless relevant to the problem, the complex $N_s d_3$ will be regarded as a variable. This is possible because N_s is to be found in the formula $Be = c_w N_s (d_3/d_1)(h_3/d_1)$.

1.   *Relevancy List*

$$\{D_{ax} \qquad w, n, \qquad d_1, d_2, N_s, d_3, d_w, d_{sp}, h_0, h_1, h_2, a_1, \qquad \rho, \eta\} \qquad (6.23)$$
$$/\text{---}/ \qquad /\text{----}/ \qquad /\text{--}/ \qquad /\text{----}/$$

| Variable | Functional parameter | Geometrical parameter | Material constants |

2.   *Dimension Matrix*

	Central matrix			Residual matrix										
	ρ	d_1	η	D_{ax}	w	n	d_2	$(N_s d_3)$	d_w	d_{sp}	h_0	h_1	h_2	a_1
M	1	0	1	0	0	0	0	0	0	0	0	0	0	0
L	-3	1	-1	2	1	0	1	1	1	1	1	1	1	1
T	0	0	-1	-1	-1	-1	0	0	0	0	0	0	0	0

M = mass, L = length, T = time.

3. *Unit Matrix*

Only four linear transformations are required to produce a unit matrix.

	Unit matrix			Residual matrix											
	ρ	d_1	η	D_{ax}	w	n	d_2	$(N_s\,d_3)$	d_w	d_{sp}	h_0	h_1	h_2	a_1
$M + T$	1	0	0	-1	-1	-1	0	0	0	0	0	0	0	0
$3M + L + 2T$	0	1	0	0	-1	-2	1	1	1	1	1	1	1	1
$-T$	0	0	1	1	1	1	0	0	0	0	0	0	0	0

It is confirmed that the rank of the matrix $r = 3$ and that $(14 - 3)$ constants exist.

The dimensionless constants are obtained according to the following rule: every quantity of the remaining matrix appears as the multiple of a fraction whose denominator is made up of the quantities of the unit matrix raised to the power indicated in the residual matrix.

4. *Determination of π Values*

$$\pi_1 = \frac{D_{ax}}{\rho^{-1}\eta} = \frac{D_{ax}}{\nu} \qquad \pi_7 = \frac{d_{sp}}{d_1} \tag{6.24}$$

$$\pi_2 = \frac{w}{\rho^{-1}d_1^{-1}\eta} = \frac{wd_1}{\nu} \qquad \pi_8 = \frac{h_0}{d_1} \tag{6.25}$$

$$\pi_3 = \frac{n}{\rho^{-1}d_1^{-2}\eta} = \frac{nd_1^2}{\nu} \qquad \pi_9 = \frac{h_1}{d_1}$$

$$\pi_4 = \frac{d_2}{d_1} \qquad \pi_{10} = \frac{h_2}{d_1}$$

$$\pi_5 = \frac{N_s d_3}{d_1} \qquad \pi_{11} = \frac{a_1}{d_1}$$

$$\pi_6 = \frac{d_w}{d_1}$$

The structure of the constants π_1, π_2 and π_3 are not meant to be interpreted and are unsuitable for practical application. They are representative of the relationship to other complex formulae. π_1 is recombined with the other π values so that ν is eliminated and a known constant is created.

5. *Recombination of π Values*

$$\pi_1^{-1}\pi_2\pi_8 = \frac{wh_0}{D_{ax}} \equiv Bo \tag{6.26}$$

This term is the Bodenstein number for axial diffusion (back-mixing). When the following recombination is carried out:

$$\pi_3 \pi_4^2 = \frac{nd_1^2}{v}\left(\frac{d_2}{d_1}\right)^2 = Re_n = \frac{nd_2^2}{\pi} \tag{6.27}$$

This is the Reynolds number for stirred Newtonian fluids. Furthermore

$$\pi_3 \pi_2^{-1} \pi_4 = \frac{nd_2}{w} \equiv N_{st} \tag{6.28}$$

where N_{st} is the Strouhal number. This constant is a modified version of the Strouhal number for the example treated here and relates the stirring speed to the linear flow velocity.

One obtains the complete set of π values with known dimensionless constants:

$$\left\{ Bo, Re_n, N_{st}, \frac{d_2}{d_1}, \frac{N_S d_3}{d_1}, \frac{d_w}{d_1}, \frac{d_{sp}}{d_1}, \frac{h_0}{d_1}, \frac{h_1}{d_1}, \frac{h_2}{d_1}, \frac{a_1}{d_1} \right\} \tag{6.29}$$

These calculations lead to no reduction in the number of constants but do provide values which can be interpreted in terms of flow mechanics.

For stirred systems a further correlation is provided by the power number or Newton number, Ne, where

$$Ne = \frac{P}{\rho n^3 d_2^5} = f\left(Re, Fr, \frac{d_2}{d_1}, \frac{h_{2m}}{d_1}, \frac{h_0}{d_1}, \frac{e}{d_1}, \text{stirrer system} \right) \tag{6.30}$$

is valid. For

$$Re_n > 10^3 - 10^4 \quad \text{at BW} > 0.2$$
$$Re_n > 10^5 \qquad \text{at BW} = 0.01 - 0.02$$

the influence of Re_n and Fr is negligible in Eqn (6.30), therefore Re_n can be removed from Eqn (6.29).

Under normal conditions the simple forms d_w/d_1, h_1/d_1, h_2/d_1, a_1/d_1 have a negligible influence on back-mixing; therefore, Eqn (6.29) can be reduced finally to

$$\left\{ B_o, N_{st}, \frac{d_2}{d_1}, \frac{N_S d_3}{d_1}, \frac{d_{sp}}{d_1}, \frac{h_0}{d_1} \right\} = 0 \tag{6.31}$$

References

ATKINSON B. and MAVITUNA, F. (1983) *Biochemical engineering and biotechnology handbook* (New York: The Nature Press).

BOX, G.E.P. and HUNTER, W.G. (1965) Sequential design of experiments for nonlinear models. *Proc. IBM Scientific Computing Symp. on Statistics, New York.*

BOX, G.E.P. and WILSON K.B. (1951) On the experimental attainment of optimum conditions. *J. Roy. Stat. Soc., Ser. B.* **13**, 1–45.

BUCKINGHAM, E. (1914) On physically similar systems, illustration of the use of dimensional equations. *Phys. Rev.* **4**, 345–76.

DIEKMANN, H. and METZ, H. (1991) *Grundlagen und Praxis der Biotechnologie* (Stuttgart: Gustav Fischer Verlag).

Methods in biotechnology

HAMAD, S.H. (1986) Screening of yeasts associated with food from the Sudan and their possible application for single cell protein and ethanol production. Doctoral thesis, Technische Universität, Berlin.

HARTMANN, K., LEZKI, E. and SCHÄFER, W. (1974) *Statistische Versuchsplanung und -auswertung in der Stoffwirtschaft* (Leipzig: Deutscher Verlag für Grundstoffindustrie).

KÖRNER, S., WOLF, K.-H. and PECINA, H. (1991) Ein Beitrag zur Modellierung des Wachstums von Basidiomyceten bei der Solid-State-Fermentation, Teil 1: Grundlagen und Modellansatz. *BIOforum* **14**, 346–9.

PIRT, S.J. (1975) *Principles of microbe and cell cultivation* (Oxford: Blackwell Scientific Publications).

RAYLEIGH, LORD (1915) The principle of similitude. *Nature* **95**, 66–8.

REHM, H.-J. and REED, G. (1985) *Biotechnology*, Vol. 2: *Fundamentals of biochemical engineering* (Weinheim: VCH).

WOLF, K.-H. (1991) *Kinetik in der Bioverfahrenstechnik* (Hamburg: B. Behr's Verlag).

WOLF, K.-H. and VOIGT, F. (1992) Einbeziehung der lag-Phase und des Temperatureinflusses in die logistiche Gleichung. *Wiss. Z. Techn. Univ. Dresden* **41**(6), 69–78.

ZLOKARNIK, M. (1991) *Dimensional analysis and scale up in chemical engineering* (Berlin: Springer Verlag).

Index